T0213814

SpringerBriefs in Applied Sciences and Technology

Manufacturing and Surface Engineering

Series editor

Joao Paulo Davim, Aveiro, Portugal

More information about this series at http://www.springer.com/series/10623

Nanjappan Natarajan · Vijayan Krishnaraj
J. Paulo Davim

Metal Matrix Composites

Synthesis, Wear Characteristics,
Machinability Study of MMC Brake Drum

 Springer

Nanjappan Natarajan
Department of Mechanical Engineering
Sri Ranganathar Institute of Engineering
  and Technology
Coimbatore
Tamil Nadu
India

J. Paulo Davim
Department of Mechanical Engineering
University of Aveiro
Aveiro
Portugal

Vijayan Krishnaraj
Department of Production Engineering
PSG College of Technology
Coimbatore
Tamil Nadu
India

ISSN 2191-530X   ISSN 2191-5318   (electronic)
ISBN 978-3-319-02984-9   ISBN 978-3-319-02985-6   (eBook)
DOI 10.1007/978-3-319-02985-6

Library of Congress Control Number: 2014946431

Springer Cham Heidelberg New York Dordrecht London

Printed on acid-free paper

Springer is part of Springer Science+Business Media (www.springer.com)

Acknowledgments

The authors Natarajan Nanjappn et al. would like to thank Elsevier Publisher for granting permission for re use of the published materials. The authors Vijayan Krishnaraj et al. would like to thank Nova Publisher for granting permission for reuse of the published materials.

Contents

Symbols

a	Deceleration (m/s^2)
A	Inside surface area of brake drum (m^2)
A_{mcp}	Area of the master cylinder (m^2)
A_s	Outer surface area of the drum (m^2)
A_{wc}	Area of the wheel cylinder (m^2)
BF	Brake factor
c_d	Specific heat of brake drum (Nm/kg °C)
c_s	Specific heat of shoe (Nm/kg °C)
D	Outer diameter of brake drum (mm)
d	Stopping distance (m)
d_c	Minimum stopping distance (m)
E	Modulus of Elasticity (GPa)
E_b	Braking energy (Nm)
F	Pedal force (N)
F_{bf}	Brake force at front wheel (N)
F_{br}	Brake force at rear wheel (N)
F_{bt}	Total brake force (N)
F_N	Normal force in the drum (N)
h_r	Convective heat transfer coefficient (Nm/h °C m^2)
I	Mass moment of inertia of rotating parts (kg-m^2)
k	Correction factor for rotating masses
k_d	Thermal conductivity of brake drum (Nm/hm^2 K)
k_s	Thermal conductivity of shoe lining (Nm/hm^2 K)
L	Inner width of brake drum (mm)
L_1	Total width of drum (mm)
m	Maximum mass of the vehicle (kg)
N	Speed of brake drums (rpm)
n_a	Number of brake applications
p	Brake line pressure (N/mm^2)
P''_{max}	Maximum brake power for one brake drum (Nm/s)
P_b	Brake power absorbed (Nm/s)

P_{bavg}	Average brake power (Nm/s)
P_{max}	Maximum brake power (Nm/s)
p_{max}	Maximum pressure on the brake drum (N/mm^2)
p_N	Average normal pressure (N/mm^2)
p_o	Push out pressure (N/mm^2)
p_θ	Pressure on the brake drum at an angle 'θ' (N/mm^2)
$q_{(o)}$	Brake power absorbed by the drum (Nm/h)
$q''_{(o)}$	Maximum heat flux into the drum (Nm/hm^2)
q''_{max}	Maximum heat flux into the drum (Nm/hm^2)
q_d	Heat flux into the brake drum (Nm/hm^2)
q_{rad}	Heat transfer through radiation (Nm/hm^2)
q_s	Heat flux into brake shoe (Nm/hm^2)
r	Radius of brake disc (mm)
R	Radius of tire (m)
R_1	Inner radius of the brake drum (mm)
R_2	Outer radius of drum (mm)
r_d	Radius of brake drum (mm)
R_d	Thermal resistance to conductive heat flow into drum
R_e	Reynold's number
r_l	Pedal lever ratio
R_s	Thermal resistance to conductive heat flow into shoe (hmK/Nm)
s	Tire slip
S_b	Stress in brake drum (N/mm^2)
t	Braking time (s)
t_b	Heat penetration time (s)
t_c	Cooling cycle time (s)
T_d	Inside temperature of brake drum (°C)
T_i	Initial temperature of brake drum (°C)
T_{max}	Maximum surface temperature rise (°C)
T_o	Outside temperature of brake drum (°C)
t_s	Time to stop the vehicle (s)
T_α	Ambient temperature (°C)
V	Velocity of vehicle (m/s)
$V_{(t)}$	Velocity at time 't' (m/s)
V_1	Velocity at begin of braking (m/s)
V_2	Velocity at end of braking (m/s)
v_d	Volume of the brake drum (m^3)
W	Maximum weight of the vehicle (N)
Z	Thickness of the brake drum (mm)
ζ_t	Temperature of brake drum at time 't' (°C)
ΔT	Change in temperature (°C)
ρ_a	Density of air (kg/m^3)
ρ_s	Density of shoe lining (kg/m^3)
ρ_d	Density of brake drum (kg/m^3)
α	Thermal diffusivity (m^2/h)

ω_1	Angular velocity of rotating parts at begin of braking (1/s)
ω_2	Angular velocity of rotating parts at begin of braking (1/s)
μ_r	Adherence coefficient between tire and road
θ	Angle subtended by a brake shoe lining
η_l	Pedal lever efficiency
η_c	Wheel cylinder efficiency
μ	Friction coefficient between brake drum and shoe
μ_a	Viscosity of air (kg/ms)
σ	Stephan Boltzman constant
ε	Emissivity of brake drum
γ	Relative brake power absorbed by the brake drum
ξ	Deformation of brake drum (mm)
σ_c	Circumferential stress in the brake drum (Pa)
σ_r	Radial stress in the brake drum (Pa)
υ	Poisson's ratio

Chapter 1
Introduction

Conventional monolithic materials have limitations in achieving good combination of strength, stiffness, toughness and density. To overcome these shortcomings and to meet the ever increasing demand of modern day technology, composites are most promising materials of recent interest. A composite is a structural material, which consist of combining two or more constituents in order to obtain a combination of properties that cannot be achieved with any of the constituents acting alone. The constituents are combined at a macroscopic level and or not soluble in each other. The constituents as well as the interface between them are recognizable and it is the behavior and properties of the interface that generally control the properties of the composite. The main difference between composite and an alloy is, in composites constituents materials are insoluble in each other and the individual constituents retain those properties, where as in alloys constituents materials are soluble in each other and form a new material which has different properties from their constituents.

The discontinuous phase in composites is usually harder and stronger than the continuous phase and is called the reinforcing agents. The continuous phase is called as the matrix. Both the reinforcements and the matrix retain those physical and chemical identities, but produce a combination of properties that cannot be achieved with either of the constituents acting alone. In general, reinforcements are the principle load carrying members, while the surrounding matrix keeps them in the desired location and orientation, acts as a load transfer medium between them and protects them from environmental damages.

Properties of composites are strongly influenced by the properties of their constituent materials, their type, their distribution and the interaction between them. Like conventional materials, composites are not homogeneous and isotropic. Composites are completely elastic up to failure; exhibit no yield point or a region of plasticity.

© The Author(s) 2015
N. Natarajan et al., *Metal Matrix Composites*, SpringerBriefs in Manufacturing and Surface Engineering, DOI 10.1007/978-3-319-02985-6_1

1.1 Metal Matrix Composites

Metal Matrix Composites are composed of a metallic matrix such as aluminium, magnesium, iron, cobalt, copper and a dispersed ceramic like oxides, carbides or metallic (lead, tungsten, molybdenum) phase. Among the various matrix materials, aluminium alloys are well suited because of light weight, environmental resistance and useful mechanical properties such as specific modulus, strength, toughness and impact resistance. Also the melting point of aluminium is high enough to assure many application requirements, sufficiently low for convenient composite processing. Silicon carbide is one of the widely used reinforcements because of its high modulus and strength, excellent thermal resistance, good corrosion resistance, good compatibility with the aluminium matrix, low cost and ready availability. The ceramic particles reinforced aluminum composites are termed as new generation material and these can be tailored and engineered with specific required properties for specific application requirements. Particle reinforced composites have a better plastic forming capability than that of the whisker or fiber reinforced ones, and thus they have emerged as most sought after material with cost advantage and they are also known for excellent heat and wear resistance applications.

1.2 Advantages of MMC

The advantages of metal matrix composites are:

- Low coefficients of thermal expansion.
- Lower creep rate.
- Better fatigue resistance.
- High performance.
- Durability.
- Excellent strength-to-weight ratio.
- Better wear resistance.
- Better radiation resistance.

1.3 Limitations of MMC

The limitations of metal matrix composites are

- Difficult for machining.
- Costlier hence supply constraints.
- Potential for environmental degradation.
- Consistency of material properties.

1.4 Applications of Composites

The applications of composites are:

- Aircrafts—rudders, elevators, landing gear doors, panels and flooring of airplanes.
- Space—payload bay doors, remote manipulator arm, high gain antenna, antenna ribs and struts.
- Marine—propeller vanes, fans and blowers, gear cases, valves and strainers and condenser shells.
- Automotive—clutch plates, engine blocks, push rods, frames, piston rods, valve guides, automotive racing brakes, drive shafts, rocker arm covers and suspension arms.
- Chemical industries—composite vessels for liquid natural gas for alternative fuel vehicle, racked bottles for fire service and storage tanks.
- Construction—tunnel supports, airport facilities such as runways and aprons, roads and bridge structures, concrete slabs, power plant facilities, marine and off-shore structures.

Thus, considerable research in the field of material science has been directed towards the development of new light-weight, high performance engineering metal matrix composites because of its useful mechanical properties such as specific modulus, strength, toughness and impact resistance.

Chapter 2
Literature Review

Applications of metal matrix composites in defense, aerospace and light vehicles have been reported by Rittner (2001). She has concluded that the scope for MMC in all the above areas were optimistic and suggested further improvement in processes, selection of alloy, selection of reinforcement and selection of components to reduce the cost of end product. Robert (2001) has presented various forms of aluminium alloys and their applications. Based on his survey on the growth of aluminium alloys, he concluded that 32.2 % of the aluminum was consumed in transport industry in different forms. Foltz and Charles (1991) have presented various matrix alloys, reinforcements and their applications in space, defense, automotive and electronic packaging. They also presented the possible applications of MMCs in making automotive components like pistons, cylinder sleeve, connecting rod and brake discs. Many Researchers (Suresh et al. 1993; Kevorkijan 1999; Rohatgi 1991; Nakanishi et al. 2002) have presented the applications of MMCs for the automotive components and the feasibility of manufacturing these materials. Surappa (2003) has presented an overview of aluminium matrix composite material systems on aspects relating to processing, microstructure, properties and applications. Many challenges of using the metal matrix composites are producing high quality and low cost reinforcements, developing simple economical and portable non-destructive kits to quantify undesirable defects, developing less expensive tools for machining and cutting and also developing re-cycling technology. The following chapters discuss the issues in design and manufacturing of an automobile brake drum.

2.1 Automotive Brake System

The brake system is the most important system in vehicles (Fig. 2.1). It converts the kinetic energy of the moving vehicle into thermal energy while stopping. The basic functions of a brake system are to slow the speed of the vehicle, to maintain

© The Author(s) 2015
N. Natarajan et al., *Metal Matrix Composites*, SpringerBriefs in Manufacturing and Surface Engineering, DOI 10.1007/978-3-319-02985-6_2

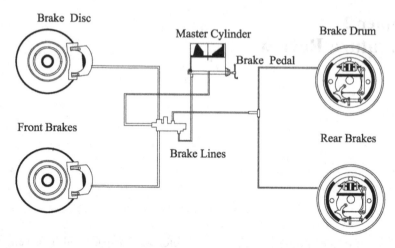

Fig. 2.1 Automotive brake system

its speed during downhill operation, and to hold the vehicle stationary after it has
come to a complete stop. The brake system is composed of master cylinder, brake
lines, wheel cylinders or slave cylinders, shoes or pads, drum or disc and brake
fluid the master cylinder is located under the hood and it is directly connected to
the pedal. It converts the foot's mechanical pressure into hydraulic pressure. A
mater cylinder has two complete separate cylinders in one housing, each handling
two wheels. Even if one cylinder fails, the other cylinder will stop the vehicle. The
brake fluid travels from the master cylinder to the wheels through a series of steel
tubes. It uses non-corrosive seamless steel tubing with special fittings at all
attachment points. Wheel cylinders are cylinders in which the movable pistons
convert the hydraulic pressure of the brake fluid into mechanical force. It consists
of a cylinder that has two pistons, one on each side. Each piston has a rubber seal
and a shaft that connect the piston with a brake shoe. The wheel cylinders of the
brake drum are made up of a cylindrical casting, an internal compression spring,
two pistons, two rubber cups or seals, and two rubber boots to prevent the entry of
dirt and water. The wheel cylinders are fitted with push rods that extent from the
outer side of each piston through a rubber boots, where they bear against the brake
shoes. Hydraulic pressure forces the pistons in the wheel cylinder which forces the
brake shoes or pads against the machined surface of the brake drums or rotors.
When the brake pedal is depressed, it moves the pistons within the master cylinder,
pressurizing the brake fluid in the brake lines and slave cylinders at each wheel.
The fluid pressure causes the wheel cylinders pistons to move, which forces the
shoes against the brake drums. Brake drums use return springs to pull the pistons
back away from the drum when the pressure is released. The brake shoes consist of
a steel shoe with the friction material or lining materials are riveted or bonded to it.
The lining materials are either asbestos (organic), semi-metallic, or asbestos free
materials. The lining material consists of fibers, fillers, binders and friction mod-
ifiers. The brake drums are made up of cast iron and have a machined surface

inside the drum where the shoes make contact. The brake drums will show the signs of wear as the lining seats themselves against the machined surface of the brake drum. When new drums are installed, the brake drum should be machined smooth. The brake fluid is special oil that has specific properties. It is designed to withstand cold temperatures without thickening as well as very high temperatures without boiling.

2.2 Materials Used in Automotive Brake Drum

2.2.1 Cast Iron

Cast iron is normally used for making the brake rotors and drums. The excellent heat absorbing capacity, low cost, simple manufacturing methods are the reasons for using the cast iron in these applications. Low corrosion resistance, rusting, brake noise, and high density are the disadvantages of using this material for brake applications. Since the brake rotors/drums represent unsprung rotating weights, increase in their mass will also increase the inertia of the rotating parts and can also decrease vehicle dynamics and acceleration. Gray cast iron with type-A graphite flakes with a pearlitic matrix of low ferrite and carbon content is used.

2.2.2 Compacted Graphite Iron

In recent years, the truck industry was in need of a lightweight brake drum. The result is a new lightweight brake drum with higher strength to be cast is Compacted Graphite Iron (CGI). Specifically because of CGI, they were able to enter a market in which neither had participated before. Designing a brake drum with cast CGI resulted in a wheel component that had high heat transfer, long life, low wear and reduced weight. CGI combines high strength with reduced weight when compared to iron or steel brake drums. It can reduce a casting's weight by 10–25 %. Its higher strength allows for thinner sections because the brake drum can withstand the loads applied. CGI has sufficient heat flow to move the heat from the brake shoe area. Cueva et al. (2003) studied and compared the wear resistance of three different types of gray cast iron (grade 250, high carbon gray iron and titanium alloyed gray iron) used in brake rotors and compared with the results obtained with a compact graphite iron (CGI). Based on the investigations they concluded that the compact graphite iron more wear higher frictional forces and temperatures.

The shape of the graphite flakes found in CGI and the metal's pearlite/ferrite matrix ratio determines its mechanical and physical properties. Figure 2.2 shows the rounded edges of graphite within CGI suppress cracking that would occur with

Fig. 2.2 Microstructure of
C. G. iron

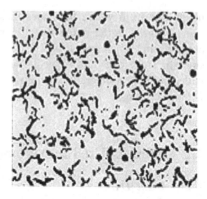

Fig. 2.3 A typical CGI brake
drum

sharp-flaked edges typically found in gray iron. The CGI brake is shown in
Fig. 2.3. These factors increase the tensile strength to 450 MPa relative to gray
iron (276 MPa). The modulus of elasticity of CGI varies from 138–165 GPa. The
variations result from differences in graphite shape and amount, section size and
matrix structure. The elastic modulus of dynamically loaded CGI components may
be 50–75 % higher than identically designed gray iron castings. The increased
modulus is equals increased stiffness.

2.2.3 Steel Shell Cast Iron Composite Brake Drum

The protective steel shell surrounding the cast iron brake surface affords major
weight savings as well as safety advantages. The steel shell developed allows
eliminating the heavy full cast iron design to keep cast drums from breaking. It
absorbs energy by converting friction between the drum surface and lining into
heat. Heat must be stored and then quickly dissipated to prevent the loss of
stopping power as well as drum and lining deterioration. The unique ribbed steel
shell provides not only strength, but also additional surface area to dissipate heat.
It also reduces excessive cast iron mass. Full cast drums must be provided with

Fig. 2.4 Steel shell cast iron composite drum

additional iron in the drum band and mounting areas in order to achieve the same strength. The steel drum back contributes to significant weight savings that result in increased payloads and improved fuel economy. The composite drum is developed out of a need to reduce the weight of passenger cars and light trucks.

The composite drum is produced by stamping a steel drum back and placing it into a mold at the foundry. Molten iron is poured into the mold and fills the cavity to take the shape of the drum pattern. As the molten iron solidifies, the steel drum back fuses with the cast iron at the edges. The steel shell cast iron and its cross section are shown in Figs. 2.4 and 2.5 respectively. A mechanical bond is also formed from tabs created during the stamping process. The composite drums require less machining than full cast drums. Typically, only the brake surface and the open lateral surfaces are machined, thereby reducing cost. A weight reduction of 10–15 % over full cast is achieved.

2.2.4 Steel Cage Reinforced Cast Iron Brake Drum

The steel cage reinforced cast iron drum incorporates two rugged materials, steel and cast iron, into one composite reinforced structure that resists checking and cracking without increasing mass. Figure 2.6 shows a reinforcing steel cage that provides strength and prolongs life of the brake drum. This steel cage is comprised of hoops and crossbars for superior radial strength and additional axial stability. There is less deflection and mechanical stress, because the steel takes the brunt of the braking load. So the drum is less sensitive to heat checking that leads to cracking and eventual breakage of the drum. In highway truck/trailer brake applications where good thermal conductivity is essential, there is no substitute for

Fig. 2.5 Cross section of
brake drum

cast iron. The inner steel cage of brake drum is literally surrounded by cast iron
providing excellent heat absorption throughout the drum while improving strength
and drum life. The machined mounting face of the cast drum provides a solid
supporting foundation for the wheels of the vehicle. This rigid mounting face
assures a completely flat mating surface for positive wheel support.

2.2.5 Fly Ash Reinforced al MMC Brake Drum

The MMCs reinforced with fly ash have the potential of being cost effective, light
weight composites with good mechanical properties. These composites are

Fig. 2.6 Steel cage
reinforced drum

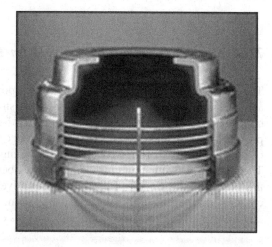

Fig. 2.7 Fly ash reinforced
brake drum

manufactured by dispersing coal fly ash in aluminium alloys. These composites offer cost saving low energy consumption and light weight. The Fig. 2.7 shows a fly ash reinforced MMC brake drum manufactured and studied in University of Wisconsin-Milwaukee. But the limitations are the segregation of low density fly ash composite, and mixing of this low density fly ash with the aluminium alloy.

2.2.6 Silicon Carbide Reinforced Copper MMC

The Metal Matrix Composites of copper alloy reinforced with silicon carbide particles are more suitable for use in brakes and other severe frictional applications because of their higher thermal conductivity, higher melting point and superior corrosion resistance. Kennedy et al. (1997) have investigated the tribological characteristics of copper based silicon carbide particulate metal matrix composites synthesized from copper coated silicon carbide particles. They performed wear tests using pin on disc method and compared the wear resistance and friction coefficient with the cast iron while sliding against friction material. They have observed that the Cu based MMCs have better wear resistance and the friction coefficient is comparable with the cast iron. The increase in wear resistance is because of the presence of the hard SiC particles in the copper matrix.

2.2.7 Thixoformed Hypereutectic Aluminium Silicon Alloy

Solidification of hypereutectic Al–Si alloys contain primary silicon particles which provides cast parts with a in situ composite structure which resembles that of a

Fig. 2.8 Cast Al–Si alloy

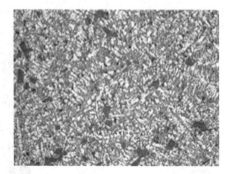

Fig. 2.9 Thixoformed alloy

Fig. 2.10 Thixoformed
brake drum

particulate MMCs reinforced with silicon inspite of silicon carbide particles as shown in Figs. 2.8 and 2.9. The resulting material has improved properties like high wear resistance, high strength and high hardness. Kapranos and Kirkwood (2003) have thixoformed an automotive brake drum using A390 hypereutectic alloy and bench tested at room temperature, 200 and 350 °C. They have found that the thixoformed drums had good thermal and wear properties when compared to cast iron drums. The drum has less weight and good wear resistance. The Fig. 2.10 shows the thixoformed brake drum.

2.2.8 Carbon/Carbon Composites

The carbon/carbon brakes are manufactured from carbons of different structural characteristics such as PAN carbon fibers and vapor deposited carbon. These composites have good thermal shock resistance, high thermal conductivity and high strength at elevated temperatures. The heat capacity is 2.5 times more than steel. The development of these materials and manufacturing technologies is underway. Blanco and Bermejo (1997) have presented the structure, properties, applications and operational behaviour of carbon-carbon disc brakes. They have investigated the wear, role of friction, dust, friction and lubrication and the mechanisms of wear. Zaidi and Senouci (1999) have studied the friction behaviour of carbon brake block/brake drum system at high sliding speed under high applied load. The contact temperature, friction behaviour and temperature evolution are determined. They have studied the mechanical and thermal damages caused under high dynamic load.

2.3 Design and Optimization of Brake Drum

Literatures are available on the general optimization procedures (Arora 1989; Rao 1995). Rajendran and Vijayarangan (2001) formulated a solution technique using genetic algorithm for design optimization of composite leaf spring. On applying the GA, they have obtained optimum dimensions for the composite leaf spring, which contributes towards achieving the minimum weight with adequate strength and stiffness. Rangaswamy and Vijayarangan (2005) have established a model to optimize the parameters of the composite drive shaft to reduce the weight of the composite drive shaft using genetic algorithm. They have achieved a considerable weight saving and the variation of the stress is also found to be within the permissible limit. Wang et al. (1999) have developed a method for shape optimization process to enable the description of the shape. Genetic algorithms are nontraditional optimization process based on the mechanics of natural genetics and natural selection of parameters. The fundamentals of genetic algorithm are explained by Raol and Jalisatgi (1996). Deb (1991) has developed the computer algorithm of GA. Goldberg (1989) has termed the operation procedure of GA as working with a coding solution, search from a population, and use probabilistic transition rules. Sandgre and Jense (1992) have suggested a new approach for the design and optimization of structural components. The permissible design space is discretized with each element assigned a design variable, which determines how it modifies the design. A generic optimization is applied to turn each element either on or off and penalty function is employed to handle design constraints stress and maximum deflection. Rajeev and Krishnamoorthy (1992) have discussed a simple genetic algorithm for optimizing structural systems with discrete design variables. Deb (1991) has used genetic algorithm to optimize welded beam structure consisting of a highly nonlinear objective function with five nonlinear constraints. Botello et al. (1999) have used genetic algorithm and simulated annealing to the optimization of pin-jointed structures. Wellman and Gemmill (1995) have applied the genetic algorithms to the performance optimization of asynchronous automatic assembly systems and the performance of genetic algorithm is measured through comparison with the results of stochastic quasi-gradient methods to the same automatic assembly systems. Duda and Jakiela (1997) have explained how the genetic algorithm is used to distribute subsets of the evolving population of solutions over the design space. Cunha et al. (1999) have presented the use of genetic algorithms as a complementary technique allowing a fist estimation of elastic coefficients. Cho and Gweon (1999) have suggested a new kind of static estimator based on genetic algorithms based on search technique.

The literature surveys carried out have shown that the optimization technique can be applied to optimize the design of components. Although many papers are available on optimization of engineering components no paper has been reported in the field of optimization of brake drum.

2.4 Manufacturing of Metal Matrix Composite

Ding et al. (2000) have designed and manufactured a front brake rotor by semi-sold stirring plus liquid forging process. Then the brake rotors are subjected to dynamometer test and the performance of the MMC brake rotor is compared with the conventional cast iron rotor. They have concluded that the MMC rotors have higher wear resistance, low temperature rise, high friction coefficient. Pai et al. (2001) have presented the low cost processing of MMCs, surface treatment of reinforcement, process parameters and the role of alloy additions with the special reference to the Al-graphite system, Al-silicon carbide, and Al-short fibers carbon systems. They have also highlighted the manufacturing of MMC components like piston rings, pistons, cylinder sleeve and connecting rods for light weight automotive applications. Pillai et al. (2001) in their investigation, they have concluded that the semisolid processing of aluminium composites have better properties like minimum interfacial reactions, uniform distribution of reinforcements and high percentage of reinforcement can be added with the matrix alloy. Degischer and Prader (2000) have presented the functions of thematic network in assessing the applications of metal matrix composite materials in all technical fields. They have also presented the role of the thematic network in sharing information on processing, testing, modeling, application and marketing of MMCs. Goni et al. (2000) have suggested that the high processing cost of MMCs, as the important barrier for using it in automotive applications. They have also suggested that the cost of MMC components can be reduced either by locally reinforcing the reinforcement or by reinforcing the MMC inserts in the required positions of the automotive components. Degischer et al. (2001) have presented the functions of thematic network in developing the processing and applications of MMCs. They have also presented the activities of the thematic network in sharing information on processing, testing, modeling, application and marketing of MMCs.

2.5 Wear Analysis

The wear and friction measurements have been proposed by various researchers (Anderson et al. 1984; Ludema 1992). Deuis et al. (1997) have presented a review on the dry sliding wear of aluminium composites. The friction and wear mechanisms of aluminium composites and the influence of applied load, sliding speed, wearing surface hardness, model for wear volume and the role of reinforcement phase are also have been presented. Tjong et al. (1997) have conducted wear tests on compo-cast aluminum silicon alloys reinforced with low volume fraction of SiC. Based on the wear test conducted on the block on ring, they have concluded that the addition low volume fraction of SiC particles is an effective way of increasing the wear resistance of the matrix alloy. Kwok and Lim (1999) have investigated the friction and wear behaviour of four Al/SiC_p composites over a wide range of sliding conditions by the use of a specially adapted high speed tester of the pin on disc configuration. Their investigations have shown that the wear rate increased with applied load, but varied in a rather complex manner with speed. The increased wear rate is due to catastrophic failures and melting of composites at higher loads and speeds.

Kwok and Lim (1999) have documented their investigations into the various mechanisms of wear as abrasive and delamination wear, a combination of abrasion, delamination, adhesion and melting. The size of reinforcement controls the high speed wear resistance of composites tested and concluded that the small SiC particles are more suitable for lower speed applications. Berns (2003) has compared the microstructures of conventional white cast irons and new metal matrix composites. Based on their instigation on wear resistance, toughness, corrosion resistance and fracture toughness, they have concluded that the replacement of MMC for the conventional cast iron in applications like ear protection components in the mining and cement industry where the abrasion by mineral grains prevails is feasible. Yang (2003) has developed a new formulation of wear coefficient and tested experimentally. He has conducted two different types of pin on disc wear tests using three commercial aluminium based metal matrix composites. Hardened steel disc is used as the sliding counterface for the MMC pins having 10, 15 and 20 % alumina reinforcements. He has derived a new wear equation based on exponential transient wear volume equation and Archard's equation and proved that they are better predictor of steady state wear coefficients.

Martin et al. (1996) have studied the influence of temperature on the wear resistance of 2618 Al alloy reinforced with 15 vol% SiC particulates and the corresponding unreinforced alloy in the temperature range 20–200 °C. They have concluded that the addition of the SiC particulates improved the wear resistance by a factor of two in the mild wear region due to the retention of mechanical properties of composites at elevated temperatures. Rohatgi et al. (1992) have developed a theoretical model to describe the steady sliding wear in a system of an alumina particle reinforced aluminium metal matrix composite pin on steel disc. The model predicts that the volumetric wear of the pin on disc system is proportional to the applied load, and depends on the particle volume fraction of the applied load, and depends on the particle volume fraction of the composite and relative hardness of the rubbing pair. They compared the theoretical results with the experimental results and found a good agreement between the experimental and observed results. Sahin (2003) has investigated the wear behaviour of aluminium alloy and its composites by means of pin-on disc type wear rig. He has carried out abrasive wear tests on 5 vol% SiCp and its matrix alloy against SiC and Al_2O_3 emery papers on a steel counter face and expressed the wear rate in terms of applied load, sliding distance and particle size using a linear factorial design approach. He has observed that composite exhibited low wear rate compared to unreinforced matrix material in both cases and conclude that the factorial design of an experiment can be successfully employed to describe the wear behaviour of aluminium alloy and its composites. Eriksson and Jacobson (2000) have studied the formation, mechanical properties and composition of the tribological surfaces such as organic brake pads sliding against grey cast iron rotor using high resolution scanning electron microscopy. They have presented the coefficient of friction, nature of contact, formation of contact plateaus. Eriksson et al. (2002) have presented the tribological contact in cast iron discs involving dry sliding contact at higher speeds and high contact forces. They have also presented the variation of contact surface and the corresponding variation of the coefficient of friction. Ostermeyer (2003) has presented the principal wear mechanism in brake systems by introducing a dynamic model of the friction coefficient by including both friction and wear. He has presented the characteristic structure formed in the contact area, the friction coefficient based on the equilibrium of flow of birth and death of contact patches. The dependence of the friction coefficient on the temperature and the fading effect of the brake system are also presented by him.

Shorowordi et al. (2004) have investigated the velocity effects on the wear, friction and tribochemistry of aluminium metal matrix composites sliding against phenolic brake pad. reinforced with 13 vol% of SiC made by stir casting followed by hot extrusion. They have concluded that higher velocity leads to lower wear rate and lower friction coefficient for both MMCs. It is also suggested that the transfer layer on MMC acts as a protective cover and helps to reduce both wear rate and friction coefficient. Straffelini et al. (2004) has investigated the effect of load and temperature on the dry sliding behaviour of Al based MMCs sliding against friction material. The dry sliding tests are carried out in a block on disc configuration by using MMC as disc material and friction material as the block. High temperature tests are carried out at three load levels, namely 500, 650 and 800 N. Heating is carried out by means of two thermo resistances. They have concluded that for loads lower than 200 N the wear is less where as the friction coefficient is quite high, around 0.45 and for loads more than 200 N the contact temperature becomes more than 150 °C and the friction decreases and wear increases. Howell and Ball (1995) have investigated the wear mechanisms of two magnesium/silicon aluminium alloys each reinforced with 20 vol% percentage SiC particulates sliding against three makes of automobile friction material. Two of the friction linings are commonly used against cast iron rotors in automobile brake systems while the third has been specifically formulated for use against aluminium metal matrix composite brake rotors. In the conventional friction material, the wear occurs through a process of three body abrasion. In case of specifically formulated friction material for use against MMC rotor wear has been found to be low. By their investigation, they have concluded that if the structure and composition of the friction linings are arranged correctly, the wear resistance and frictional performance of aluminium MMC brake rotors are superior to those of cast iron rotors. The wear of materials at high temperature and the transfer films have been investigated by various researchers (Liu 1980, 2004; Celik et al.2005; Sallit et al.1998).

2.6 Testing of MMC for Brake Drum Applications

Valente and Billi (2001) have developed an innovative test apparatus to induce cyclic thermal stresses in MMC specimens. The modification induced and the residual properties of the specimens are then investigated through mechanical tests and fractographic analysis. Their results have showed that MMC has good resistance to thermal cycling at least at the maximum temperature of 300 °C. They have concluded that the possibility of employing this material to car braking is good and suggested further investigations. Guner et al. (2004) have developed a cost effective approach to investigate brake system parameters. They have investigated the dynamic and thermal behaviour of the braking phenomenon by establishing a dynamic model. They have used the Newmark integration scheme and predicted the stopping distance and speed and verified these using test results. They have also conducted thermal analysis and found an excellent agreement between the numerical and test results. Roberts and Day (2000) have established a new systematic approach to design-evaluation-test product development cycle wherein the vehicle design and simulation environments are integrated. By implementing virtual testing early in the product development cycle has the potential to shorten development time, reduce risk of failure during expensive physical testing, and increase the overall product quality. Limpert (1999) has developed models for determining the temperature rise during single stop, continuous and repeated braking. Ramachandra Rao et al. (1993) have simulated the temperature distribution and brake torque developed in a brake drum using finite element methods. They have observed a good agreement between the simulated results and the observations carried out using an inertia dynamometer. Ilinca et al. (2001) have determined the effect of contact area on the temperature distribution in brake drum which results in wear and deformation of brake drum.

2.7 Summary

The literature survey carried out has shown that only few papers are available about the applications of Al MMCs. No paper has been reported in the following fields for using the Al MMCs for the brake drum applications. Hence, it has been decided to carry out the above studies to evaluate the suitability of Al MMC for light weight automotive brake drum applications.

References

Anderson AE (1984) Friction and wear of automotive brakes, ASME Handbook, Wear 18:569–577

Arora JS (1989) Introduction to optimum design. Mc Graw Hill, Newyork

Berns H (2003) Comparison of wear resistant MMC and white cast iron. Wear 254:47–54

Blanco C, Bermejo J (1997) Chemical and physical properties of carbon as related to brake performance. Wear 213:1–12

Botello S et al (1999) Solving structural optimization problems with genetic algorithms and simulated annealing. Int J Numer Meth Eng 45:1069–1084

Celik O, Ahlatci H, Kayali ES, Cimenoglu H (2005) 'High temperature abrasive wear behavior of an as cast ductile iron. Wear 258:189–193

Cho DJ, Gweon DG (1999) A state estimator design for multimodal adjustment process using genetic algorithm. Mechatronics 9(2):163–183

Cueva G, Sinatora A, Guesser WL, Tschiptschin AP (2003) Wear resistance of cast irons used in brake disc rotors. Wear 255:1256–1260

Cunha J et al (1999) Application of genetic algorithms for the identification of elastic constants of composite materials from dynamic tests. Int J Numer Meth Eng 45:891–900

Deb K (1991) Optimal design of a welded beam via genetic algorithms. AIAA J 29(11):2013–2015

Degischer HP, Prader P (2000) Assessment of metal matrix composites for innovations: a thematic network within EU frame work 4. Mater Sci Technol 16(7–8):739–742

Degischer HP, Prader P, Marchi CS (2001) Assessment of metal matrix composites for innovations: intermediate report of a European thematic network. Compos Part-A 32:1161–1166

Deuis RL, Subramanian C, Yelllup JM (1997) Dry sliding wear of aluminium composites–a review. Compos Sci Technol 57:415–435

Ding ZL, Fan YC, Qi HB, Ren DL, Guo JB, Jiang ZQ (2000) Study on the SiC_p/Al—alloy composite automotive brake rotos. Acta Metallurgica Sinica (English Letters) 13(4):974–980

Duda JW, Jakiela MJ (1997) Generation and classification of structural topologies with genetic algorithm speciation. ASME J Mech Des 119:127–131

Eriksson M, Jacobson S (2000) Tribological surfaces of organic brake pads. Tribol Int 33:817–827

Eriksson M, Bergman F, Jacobson S (2002) On the nature of tribological contact in automotive brakes. Wear 252:26–36

Foltz JV, Charles M (1991) Metal matrix composites. ASM Handb 4:903–912

Goldberg DE (1989) Genetic algorithms in search, optimization and machine learning. Addison-Wesley, Reading

Goni J, Mitxelena I, Coleto J (2000) Development of low cost metal matrix composites for commercial applications. Mater Sci Technol 16:743–746

Guner R, Yavuz N, Kopmaz O, Ozturk F, Korkmaz I (2004) Validation of analytical model of vehicle brake system. Int J Veh Des 35(4):340–348

Howell GJ, Ball A (1995) Dry sliding wear of particulate-reinforced aluminium alloys against automobile friction materials. Wear 181–183:379–390

Ilinca A, Ilinca F, Falah B (2001) Numerical and analytical investigation of temperature distribution in a brake drum with simulated defects. Int J Veh Des 26(2/3):146–160

Kapranos P, Kirkwood DH (2003) Thixoforming of an automotive part in A 390 hypereutectic Al-Si Alloy. J Materi Process Technol 135:271–277

Kennedy FE, Balbahadur AC, Lashmore DS (1997) The friction and wear of Cu-based silicon carbide particulate metal matrix composites for brake applications. Wear 203–204:715–721

Kevorkijan VM (1999) Aluminium composites for automotive applications-a global perspective. JOM 51(II):54–58

Kwok JKM, Lim SC (1999) High speed tribological properties of some Al/SiC composites: I frictional and wear rate characteristics. Compos Sci Technol 59:55–63

Limpert R (1999) Brake design and safety. SEA International, Pittsburgh

Liu T, Rhee SK, Lawson KL (1980) A study of wear rate and transfer films of friction materials. Wear 60:1–12

Ludema KC (1992) Sliding and adhesive wear. ASME Handb 18:569–577

Martin A, Martinez MA, LLorca J (1996) Wear of SiC-reinforced Al-matrix composites in the temperature range 20–200 °C. Wear 193:169–179

Nakanishi H, Kakihara K, Tomiyuki AN (2002) Development of aluminium metal matrix composites (Al-MMC), Brake rotor and pad. JSEA Rev 23(3):365–370

Ostermeyer GP (2003) On the dynamics of the friction coefficient. Wear 254:852–858

Pai PC, Pillai RM, Sathyanarayana KG, Sukumaran K, Pillai UTS, Pillai SKG, Ravikumar KK (2001) Discontinuously reinforced aluminium alloy matrix composites. Met Mater Processes 13(2–4):255–278 (Meshap Science Publishers)

Pillai RM, Pai PC, Sathyanarayana KG, Kelukutty VS (2001) Discontinuously reinforced aluminium alloy matrix composites. Met Mater Processes 13(2–4):279–290 (Meshap Science Publishers)

Rajeev S, Krishnamoorthy CS (1992) Discrete optimization of structures using genetic algorithms. J. Struct Eng 118:1233–1249

Rajendran I, Vijayarangan S (2001) Optimal design of composite leaf spring using genetic algorithms. Comput Struct 79:1121–1129

Ramachandra Rao VTVS, Rajaram LS, Seetharamu KN (1993) Temperature and torque determination in brake drums. Sadhana 18:963–983 Part 6

Rangaswamy T, Vijayarangan S (2005) A genetic algorithm based optimum design and stress analysis of laminated drive shafts. Paritantra 11(1):45–53

Rao SS (1995) Optimization theory and applications. New-age International, New Delhi

Raol JR, Jalisatgi A (1996) From genetics to genetic algorithms-solution to optimization problems using natural systems. Resonance 1(1):43–54

Rittner MN (2001) Expanding world markets for MMCs. Int J Powder Metall 37(5):37–38

Robert ES (2001) Technology innovation in aluminium products. JOM 53(2):21–25

Roberts SG, Day TD (2000) Integrating design and virtual test environments for brake component design and material selection. SAE technical paper series 2000-01-1294, Warrendale, PA 15096-0001, USA

Rohatgi P (1991) Cast aluminium metal matrix composites for automotive applications. J Metall 43:10–15

Rohatgi PK, Liu Y, Ray S (1992) Friction and wear of metal matrix composites. ASM Handb 18:801–810

Sahin Y (2003) Wear behaviour of aluminium alloy and its composites reinforced by SiC particles using statistical analysis. Mater Des 24:95–103

Sallit C, Adam RR, Valloire FR (1998) Characterization methodology of a tribological couple metal matrix composite/brake pads. Mater Charact 40(3):169–188

Sandgre E, Jense E (1992) Automotive structural design employing a genetic algorithm, SAE-92077, pp 1003–1014

Shorowordi KM, Haseeb ASMA, Celis JP (2004) Velocity effects on the wear friction and tribochemistry of aluminium MMC sliding against phenolic brake pad. Wear 256:1176–1181

Straffelini G, Pellizzari M, Molinary A (2004) Influence of load and temperature on the dry sliding behaviour of Al-based metal matrix composites against friction material. Wear 256:754–763

Surappa MK (2003) Aluminium matrix composite—challenges and opportunities. Sadhana 28:319–334, Parts 1 and 2

Suresh S, Mortensen A, Needleman A (1993) Fundamentals of metal matrixcomposites. Butterworth, Heinemann

Tjong SC, Wu SQ, Liao HC (1997) Wear behaviour of an Al-12 % Si Alloy Reinforced with a low volume fraction of SiC particles. Compos Sci Technol 57:1551–1558

Valente M, Billi F (2001) Micromechanical Modification induced by cyclic thermal stress on metal matrix composites for automotive applications. Compos Part-B 32:529–533

Wang X et al (1999) A physics based parameterization method for shape optimization. Comput Methods Appl Mech Eng 175:41–51

Wellman MA, Gemmill DD (1995) Genetic algorithm approach to optimization of asynchronous automatic assembly systems. Int J Flex Manuf Syst 7:27–46

Yang LJ (2003) Wear coefficient equation for aluminium-based matrix composites against steel disc. Wear 255:579–592

Zaidi H, Senouci A (1999) Thermal tribological behaviour of composite carbon metal/steel brake. Appl Surf Sci 144–145:265–271

Chapter 3
Synthesis of Metal Matrix Composites

3.1 Introduction

Of late, the automobile industry has been facing substantial technical challenges as it seeks to improve fuel economy, reduce vehicle emissions and enhance performance. It is important to reduce the overall weight of the vehicle for improving the fuel economy. Since the brake drum represents the unsprung rotating masses, the reduction in their weight is essential to increase the vehicle dynamics and acceleration. The possibility to substitute alternate materials, in order to improve brake performance and to reduce weight has made the development of advanced materials. Increased traffic density in cities requires brakes with more energy absorbing capability at more frequent intervals. Increased speed of automobiles with a demand of fuel economy, vehicle comfort and cost reduction envisages the suitable selection of materials for brake drums. Thus the need arises for searching a suitable material and designing comparatively smaller and light weight brakes.

Cueva et al. (2003) studied and compared the wear resistance of three different types of gray cast iron (grade 250, high carbon gray iron and titanium alloyed gray iron) used in brake rotors and compared with the results obtained with a compact graphite iron (CGI). The Metal Matrix Composites of copper alloy reinforced with silicon carbide particles are more suitable for use in brakes and other severe frictional applications because of their higher thermal conductivity, higher melting point and superior corrosion resistance. Kennedy et al. (1997) have investigated the tribological characteristics of copper based silicon carbide particulate metal matrix composites synthesized from copper coated silicon carbide particles. Kapranos and Kirkwood (2003) have thixoformed an automotive brake drum using A390 hypereutectic alloy and bench tested at room temperature, 200 and 350°C. The use of light weight materials in vehicles has been increasing as the need for higher fuel efficiencies and higher vehicle performance increase. Cast iron is the traditional material used for making brake drums and rotors of both light and heavy vehicles. The wide availability and the low cost are the advantages of these

© The Author(s) 2015
N. Natarajan et al., *Metal Matrix Composites*, SpringerBriefs in Manufacturing and Surface Engineering, DOI 10.1007/978-3-319-02985-6_3

materials. Disadvantages include its heavy weight, high wear rate, noise and vibration. There has been interest in using aluminum based metal matrix composites (Al MMCs) for brake disc and drum materials in recent years. While much lighter than cast iron, they are not as resistant to high temperatures and are sometimes only used on the rear axles of automobiles because the energy dissipation requirements are not as severe compared with the front axle. Therefore, from the performance standpoint it is very much essential to study the behaviour of MMC brake drum before using it in the actual applications. In this study, an attempt has been made to study and compare the behaviour of MMC with the conventional cast iron for brake drum applications.

3.1.1 Selection of Metal Matrix Composite

There has been interest in using aluminum based metal matrix composites (MMCs) for brake disc and drum materials in recent years. While much lighter than cast iron, they are not as resistant to high temperatures and are sometimes only used on the rear axles of automobiles because the energy dissipation requirements are not as severe compared with the front axle. While applying brake, the brakes convert the kinetic energy of the moving vehicle into thermal energy. This thermal energy diffuses through conduction within the brake drum and dissipates by convection and radiation from the outer surface of the brake drum. The material used for the brake drum should have the required physical, mechanical and thermal properties apart from being light weight. The failure of the brake drum is due to the high temperature generated inside the drum and also due to the high stresses applied on the drum.

3.1.1.1 Selection of Matrix

Matrix is a continuous phase in which the reinforcement is uniformly distributed. The advantages of metallic matrices as compared to polymer matrices are their higher tensile strength, and shear modulus, high melting point, low coefficient of thermal expansion, resistance to moisture, dimensional stability, high ductility and toughness. For the matrix, characteristics such as density, strength, high temperature strength, ductility and toughness are to be considered. Generally Al, Ti, Mg, Ni, Cu, Pb, Fe, Ag, Zn have been used as the matrix material. The main focus is given to aluminium matrix because of its unique combination of good corrosion resistance and excellent mechanical properties. The combination of light weight, high thermal conductivity, and low cost makes the aluminium matrix well suited for use as a matrix metal. The melting point of aluminum is high enough to satisfy the application requirements. Also aluminium can accommodate a variety of reinforcing agents including continuous boron, aluminium oxide, SiC, and graphite fibers and various particles, short fibers and whiskers.

3.1.2 Selection of Reinforcement

Reinforcement increases strength, stiffness and temperature resistance capacities of MMCs. The reinforcement has different sizes, shapes and volume fractions in the composite. The reinforcement may be continuous fibers, whiskers or particles. The continuous fiber reinforcement has superior properties than the discontinuous reinforcements, but suffers from the disadvantage of anisotropic properties added to the need of adopting near net shape forming techniques. The discontinuous reinforcement offers isotropic properties and amenability to be processed by conventional secondary metal forming techniques. Particulates are most common and least costly reinforcement materials. The SiC particulate reinforced Al MMCs have good potential for use as wear resistance materials. Actually particulates lead to a favorable effect on properties such as hardness, wear resistance and compressive strength. Selection criteria for the ceramic reinforcement include factors like elastic modulus, tensile strength, density, melting temperature, thermal stability, compatibility with matrix material and cost effectiveness. The silicon carbide particulate is selected since it satisfies all the above requirements.

3.1.3 Properties of A356/SiCp MMC

Aluminium alloy reinforced with SiC particles exhibit increased strength and stiffness as compared to non-reinforced aluminium alloy. In contrast to the base metal, the composite retains its room temperature tensile strength at higher temperatures. Discontinuous silicon carbide/aluminium MMCs are being developed by the aerospace industry for use in airplane skins, intercoastal ribs, and electrical equipment. In the liquid metal processing technique, the molten aluminium has the tendency to react with the reinforcing materials. The severity of the reaction is based on the kinetics and the prevailing thermodynamic conditions. The presence of alloying elements in the matrix has the influence on viscosity, contact angle and reaction rate with the dispersed particles. In the case of SiC, the following reaction is observed with the molten aluminium alloy (Pai et al. 2001).

$$3SiC + 2Al \rightarrow Al_2C_3 + 3Si \qquad \Delta G = -51.3 \, kJ \, mol^{-1}$$

However, the above reaction can be prevented by the presence of about 8 % Silicon in the matrix, wherein the higher chemical potential of SiC retards reaction. In A356/LM25 cast aluminium alloys, the above reaction is not observed. So, this combination of the A356 and SiC is more suitable for making MMCs with good mechanical properties. The mechanical properties of cast iron and A356/SiC is given in Table 3.1 (Limpert 1999; Foltz 1991).

Property	Cast iron	A356/SiC$_P$
Tensile strength (MPa)	276	300
Young's modulus (GPa)	100	112
Poisson's ratio	0.3	0.3
Density (kg/m^3)	7228	2828
Thermal conductivity (Nm/hm k)	174465	374400
Specific heat (Nm/kg k)	419	970

3.2 Manufacturing of MMC Brake Drum

The fabrication of manufacturing set-up and the manufacturing of MMC brake drum are explained in this chapter. The manufacturing of MMC brake drum involves the casting, heat treatment and the machining of brake drum to required size. Even though the MMCs have highly promising mechanical and thermal properties, they have been afforded only limited use in very specific applications. Shortcomings such as complex processing requirements and the high cost of the final product have been the greatest barriers for their proliferation. There are different methods available for fabricating the MMCs and are classified into any one of the following categories.

3.2.1 Solid State Process

The solid state processes are generally used to obtain the highest mechanical properties particularly in discontinuous MMCs. In these processes, segregation effects and the formation of brittle reaction product are minimum when compared to liquid state processes.

3.2.2 Powder Consolidation

Powder metallurgy is one of the most common methods for fabricating ceramic particles reinforced MMCs. Blending of metallic powder with ceramic fibers or particulate is a versatile technique for MMC production. After blending the metallic powder with ceramic reinforcement particulate, cold isostatic pressing is utilized to obtain a green compact that is then thoroughly outgassed and forged or extruded.

3.2.3 Diffusion Bonding

Diffusion bonding is used for consolidating alternate layers of foils and fibers to create single ply or multiply composites. This is a solid state creep deformation process. The creep flow of the matrix between the fibers make complete metal to metal contact, diffusion across the foil interfaces completes the process. The pressure and the time requirements for consolidation can be determined from matrix flow stress, taking into consideration of the above matrix flow processes. To avoid fiber degradation, care must be exercised to maintain low pressure during consolidation.

3.2.4 Liquid State Process

A majority of commercially viable applications are now produced by liquid state processing because of certain inherent advantages of this processing technique. The liquid metal is generally less expensive and easier to handle than powders. Also the composites can be produced in a wide variety of shapes, making use of different methods already developed in the casting industry for non-reinforced metals. Liquid state processing technologies are currently being investigated and developed utilize a variety of methods to physically combine the matrix and the reinforcement. The important manufacturing processes are infiltration, dispersion, spraying and in situ.

3.2.5 Infiltration

Infiltration processes involve holding a porous body of the reinforcing phase within a mould and infiltrating it with molten metal that flows through interstices to fill the pores and produce a composite. The infiltration process and the pressure driven infiltration process are shown in Figs. 3.1 and 3.2 respectively. The main parameters in infiltration processes are the initial composition, morphology, volume fraction and temperature of reinforcement, the initial composition, and temperature of infiltrating metal, and the nature and magnitude and external force applied to the metal. In some cases, the metals spontaneously infiltrate the reinforcement without any external force applied on it. To overcome poor wetting, mechanical work is applied on the molten metal to force the molten metal into the preform. The pressure required for infiltration can readily be calculated on the basis of the necessary meniscus curvature and corrections can be made for melt/fiber wetting. In practice, substantial pressures in the MPa range are likely to be needed. In most cases, fibers do not act as preferential crystal nucleation sites during melt solidification.

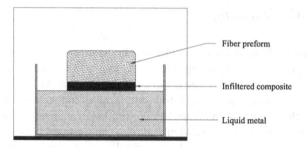

Fig. 3.1 Infiltration process

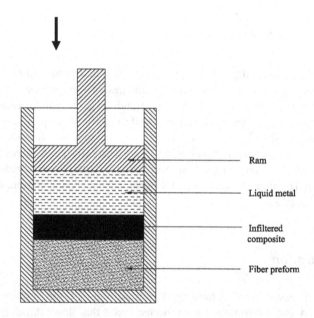

Fig. 3.2 Pressure driven infiltration process

One consequence of this is that the last liquid to freeze, which is normally solute-enriched, tends to be located around the fibers. Such prolonged fiber/melt contact, often under high hydrostatic pressure and with solute enrichment, tends to favour formation of a strong interfacial bond.

3.2.6 Dispersion Process

In dispersion process, the reinforcement is incorporated in loose form into the metal matrix. Since most reinforcement system exhibit poor wetting, mechanical force is

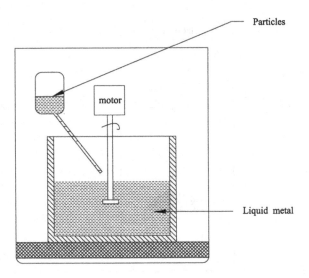

Particles

motor

Liquid metal

Fig. 3.3 Dispersion process

required to combine the phases, generally through stirring. This method is currently the most inexpensive method to produce MMCs. MMCs can be produced in large quantities which can be further processed through conventional manufacturing processes. Figure 3.3 shows the dispersion process. The simplest method is the Vortex method, which consists of vigorous stirring of liquid metal and the addition of particles in the vortex. This method usually involves prolonged liquid-ceramic contact, which can cause substantial interfacial reaction. This has been studied in detail for Al-SiC, in which the formation of Al4 C3 and Si can be extensive. This both degrades the final properties of the composite and raises the viscosity of the slurry, making subsequent casting difficult. The rate of reaction is reduced, and can become zero, if the melt is Si-rich, either by prior alloying or as a result of the reaction.

The reaction kinetics and Si levels needed to eliminate it are such that it has been concluded that the casting of Al-SiCp involving prolonged melt holding operations is suited to conventional (high Silicon) casting alloys, but not to most wrought alloys. Porosity resulting from gas entrapment during mixing, oxide inclusions, reaction between metal and reinforcement, particle migration and clustering are the critical factors for the success of this dispersion process.

3.2.7 Spray Process

In this process, droplets of molten metal are sprayed together with the reinforcing phase and collected on a substrate where metal solidification is completed. The critical parameters in spray processing are the initial temperature, size distribution and velocity of the metal drops, the velocity, temperature and feeding rate of the

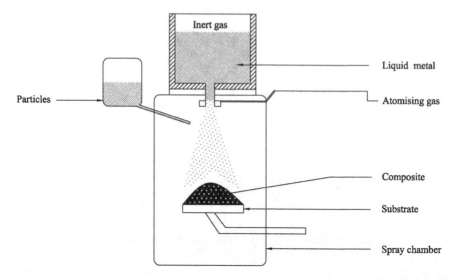

Fig. 3.4 Spray process

reinforcement and the nature and temperature of the substrate collecting the material. Most spray processes use gasses to atomize the molten metal into fine droplet stream. One advantage of the spray process is the fine grain size and low segregation of the resulting matrix microstructures. Figure 3.4 shows the spray process.

3.2.8 In Situ Process

The in situ composites are first used for the materials produced through solidification of polyphase alloys. When the polyphase alloys solidify, they may exhibit a fine lamellar or rod like structure of β phase in an α phase matrix. The reinforced inter-metallic alloys may be produced by controlled solidification or by chemical reaction between a melt and solid or gaseous phases. The schematic diagram of manufacturing in situ composites through the reaction of molten metal with a gas is shown in Fig. 3.5.

3.2.9 Deposition Process

In a typical low pressure plasma deposition process, the aluminium (alloy) powder plus reinforcement are fed into low pressure plasma. In the plasma, the matrix is heated above its melting point and accelerated by fast moving plasma gasses. These droplets are then projected on a substrate, together with the reinforcement particles. The latter particles remain solid during the whole process if one use

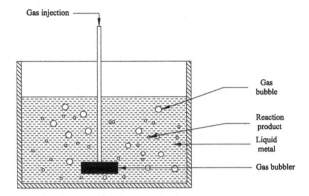

Fig. 3.5 In-situ process

lower power settings or may be partially or fully melted when higher power settings are used. By a gradual change of the feeding powder composition, gradient materials can easily be produced.

3.3 Fabrication of Dispersion Process Casting Set-up

A dispersion process casting setup has been fabricated for manufacturing the Al MMC. It consists of three basic arrangements as shown in Fig. 3.6. The melting arrangement has been used for melting and holding the liquid metal at a desired temperature (750 °C) while stirring.

The stirring arrangement has been designed to create a vortex on the molten metal. The schematic representation of the dispersion casting set-up is shown in Fig. 3.6.

3.3.1 Melting Arrangement

An electric resistance type furnace with a temperature controller has been used to melt the aluminium alloy. The furnace has a capacity of 3.5 kW power rating with a maximum operating temperature of 1200 °C and capable of melting 5 kg of aluminium in a batch. The furnace has the provision to maintain the required temperature. A graphite crucible has been used to hold the aluminium alloy inside the furnace. It has a provision to take it and pour the molten metal into the mold.

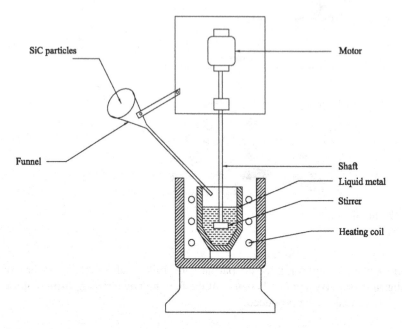

Fig. 3.6 Schematic diagram of the dispersion casting set-up

3.3.2 Feeding Arrangement

The feeding arrangement facilitates the uniform feeding of the preheated reinforcement particles into the vortex of the liquid metal. A funnel fitted with a metal tube has been fixed along the stirring arrangement. The free end of the metal tube is kept just above the vortex of the molten metal when the stirring setup is at the lowest position.

3.3.3 Fabrication of Stirrer Arrangement

A stirrer, which has four blades welded perpendicular to each other on one end of the long steel rod, has been fixed to the shaft of the motor. A sliding arrangement has been made for lowering and rising of the motor and stirring assembly. The electric motor has been connected to a speed controller (autotransformer) to vary the speed of the stirrer.

3.4 Manufacturing of Al MMC Brake Drum

The MMC brake drum has been manufactured through dispersion process. The sequence of manufacturing consists of melting the aluminium alloy, preheating the reinforcement particles, creating vortex on the molten metal through stirring, uniform feeding of the reinforcements and the pouring of the composite slurry into the mould.

3.4.1 Melting of Aluminum Alloy

The required quantity (3 kg) of A356 aluminium alloy has been placed in the graphite crucible and kept inside the furnace. The alloy is heated in the electrical resistance furnace and the temperature is set at 750 °C using the temperature controller. The temperature of the aluminium alloy has been increased and the melting takes place at 660 °C. Then, the temperature of the molten increases and it is maintained at 750 °C. The melting of the aluminium alloy is shown in Fig. 3.9.

3.4.2 Preheating of Reinforcement Particle

The preheating of the reinforcement particles is carried out in an electrical muffle furnace. The reinforcement is heated up to a temperature of 750 °C and maintained at this temperature for 30 min.

3.4.3 Stirring

The motor-stirrer assembly is lowered so that the stirrer is submerged in the molten metal. Then, the stirrer is rotated at a speed of 500 rpm using the speed controller. This creates a vortex on the molten metal surface. The preheated silicon carbide particle (43 μm) is uniformly fed through the feeding arrangement. The feeding and the stirring are simultaneously carried out until the calculated preheated reinforcement is fed into the molten metal. After the addition of the reinforcement, stirring is continued for 10 min at a reduced speed of 100 rpm to make the homogenous composite slurry.

Fig. 3.7 Aluminum alloy

Fig. 3.8 SiC particles

3.4.4 Casting

The composite slurry is then quickly poured into a mould and allowed to solidify. The solidified casting has been removed from the mould and cleaned. The gates and risers are removed and it is subjected to T6 heat treatment. The aluminum alloy and the reinforcement (SiC) used for making the MMC brake drum are shown in Figs. 3.7 and 3.8 respectively (Fig. 3.9).

3.4.5 Machining of Casting

The cast MMC drum is then machined in a lathe machine using the carbide tool. The machining has been found to be very tough because of the presence of the hard ceramic reinforcements. The turning and the facing operations have been completed to get the required dimensions for the brake drum. The machined MMC brake drum is shown in Fig. 3.10. Figure 3.11 show's the brake drum assembly.

Fig. 3.9 Manufacturing set-up

Fig. 3.10 MMC brake drum

3.5 Summary

A dispersion casting set-up has been fabricated to manufacture the MMC brake drum. The MMC brake drum has been manufactured by using this casting set-up. The cast MMC brake drum is then machined to the required size using a conventional lathe machine using a diamond tipped tool. A specimen from the same lot is subjected to optical microscopy to ensure the uniform distribution of the reinforcement in the matrix. It can be concluded that the dispersion process is the economical and viable method for mass production of MMCs.

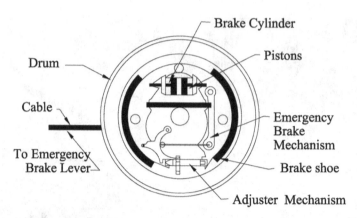

Fig. 3.11 Brake drum assembly

References

Cueva G, Sinatora A, Guesser WL, Tschiptschin AP (2003) Wear resistance of cast irons used in brake disc rotors. Wear 255:1256–1260

Foltz JV, Charles M (1991) Metal Matrix Composites. ASM Hand book 4:903–912

Kapranos P, Kirkwood DH (2003) Thixoforming of an automotive part in A 390 hypereutectic Al-Si Alloy. J Materi Process Technol 135:271–277

Kennedy FE, Balbahadur AC, Lashmore DS (1997) The friction and wear of Cu-based silicon carbide particulate metal matrix composites for brake applications. Wear 203–204:715–721

Limpert R (1999) Brake Design and Safety, SEA International 2nd edn. Warrendale

Pai PC, Pillai RM, Sathyanarayana KG, Sukumaran K, Pillai UTS, Pillai SKG, Ravikumar KK (2001) 'Discontinuously reinforced aluminium alloy matrix composites. Metals Materials and Processes, Meshap Science Publishers, Vol 13, No. 2–4, pp 255–278

Chapter 4
Fabrication of Experimental Set-up and Testing

4.1 Introduction

Guner et al. (2004) have developed a cost effective approach to investigate brake system parameters. They have Investigated the dynamic and thermal behavior of the braking phenomenon by establishing a dynamic model. Roberts and Day (2000) have established a new systematic approach to design-evaluation-test product development cycle wherein the vehicle design and simulation environments are integrated. Limpert (1999) has developed models for determining the temperature rise during single stop, continuous and repeated braking. Ramachandra Rao et al. (1993) have simulated the temperature distribution and brake torque developed in a brake drum using finite element methods. They have observed a good agreement between the simulated results and the observations carried out using an inertia dynamometer. Ilinca et al. (2001) have determined the effect of contact area on the temperature distribution in brake drum which results in wear and deformation of brake drum. The fabrication of the experimental set-up and the testing of cast iron and the Al MMC brake drums have been reported in this chapter. The test rig consists of arrangements for rotating the brake drum at different speeds, applying brake force, measuring the temperature rise in the brake drum and measuring the brake torque absorbed by the brake drum. The thermal analysis under single, continuous and repeated braking has been also presented. During braking, the kinetic energy of vehicle is converted into thermal energy due to friction at the contact surfaces. Heat generation increases the temperature of the brake components and the tire which can cause brake failure leading to accidents. The dynamic friction coefficient of cast iron and metal matrix composite brake drum has been experimentally determined. Dynamic models have been developed to predict the temperature rise in single, repeated and continuous braking. The frictional force is substituted in the equation of motion to predict the deceleration at different brake line pressures. The conventional grey cast iron brake drum and the Al MMC brake drum have been tested and the results are compared with the

© The Author(s) 2015
N. Natarajan et al., *Metal Matrix Composites*, SpringerBriefs in Manufacturing and Surface Engineering, DOI 10.1007/978-3-319-02985-6_4

analytical results. Good agreement has been observed between the analytical and the experimental results. The performance of the aluminium matrix composite brake drum has been observed to be optimistic for using it in the passenger car.

4.1.1 Power Transmission in Brake System

The primary components of a hydraulic brake system are master cylinder, limit valve and wheel cylinders as shown in Fig. 2.1. When the brake pedal is actuated, the pressure from the master cylinder is transmitted to the wheel cylinders through limit valve. The diameter of the master cylinder and the wheel cylinders are designed in such a way that the force is multiplied. The wheel pistons force the friction lining against the rotating brake drum. There exists a definite ratio between the applied brake force and the frictional force developed in the brake drum and is known as brake factor (BF). Due to friction, the brake forces F_{bf} and F_{br} are obtained on the front and rear wheels in a direction opposite direction of rotation. The dynamic behaviour of the brake system is modeled for the purpose of dynamic and thermal analysis of brake system.

4.1.1.1 Analysis of Hydraulic Brake System

The hydraulic pressure produced by the pedal actuation is obtained as

$$p = \frac{F r_L \eta_L}{A_{mcp}} \qquad (4.1)$$

The brake force exerted at the front axle is obtained as

$$F_{bf} = 2\left[(p - p_o)A_{wc}\eta_c BF\left(\frac{r_r}{R_T}\right)\right]_f \qquad (4.2)$$

The brake force absorbed at the rear axle is expressed as

$$F_{br} = 2\left[(p - p_o)A_{wc}\eta_c BF\left(\frac{r_d}{R_T}\right)\right]_r \qquad (4.3)$$

The total brake force absorbed due to applied pressure is given by

$$F_{bt} = F_{bf} + F_{br} \qquad (4.4)$$

The total brake force applied on each axle reduces the speed of rotation due to friction. The deceleration of the vehicle when the wheel is in unlocked condition is expressed as

$$a = \frac{F_{bt}}{m} \tag{4.5}$$

The time required to stop the vehicle is expressed as

$$t_b = \frac{V_i}{a} \tag{4.6}$$

The distance traveled by the vehicle before stopping due to the applied force is expressed as

$$d = V_i t_b - \frac{a t_b^2}{2} \tag{4.7}$$

The minimum stopping distance (d_c) required to stop a vehicle depends on the adherence coefficient between the tire and the road and is expressed as

$$d_c = \frac{V_i^2}{2 \mu_r g} \tag{4.8}$$

4.1.1.2 Brake Power Absorbed in a Brake Drum

Brake drum is a structural component that converts the kinetic energy of a moving vehicle into thermal energy in the process of slowing down or stopping of a vehicle. Due to the friction between the stationary brake shoe lining and the rotating brake drum, a drag force is developed in a direction opposite to the direction of rotation. Therefore, the brake power developed in a brake drum can be expressed as follows.

$$P_b = (p - p_o) A_{wc} \eta_c BF \left(\frac{r_d}{R_T} \right) V_i \tag{4.9}$$

4.1.2 Thermal Analysis of Brake Drum

During braking, the kinetic energy of the moving vehicle is converted into thermal energy through friction in the brakes. With high deceleration level resulting high heat generation in a single stop, the braking time may be less than the time required for the heat to penetrate through the drum and the shoe which will lead to temperature rise at the interface. The dimensions of brake drum used for determining the brake factor are shown in Fig. 3.11. The increase in temperature in the brake drum without ambient cooling is expressed as

$$\Delta T = \frac{P_b t_b}{\rho_d c_d v_d} \tag{4.10}$$

While applying brake, about 90 % of the thermal energy is absorbed by the brake drum and the rest is dissipated into the brake shoe based on the thermal resistance of brake drum and the shoe materials. This thermal energy diffuses through conduction inside the brake drum and dissipates by convection and radiation from the outer surface of the brake drum. The heat transfer from inside the brake drum to outside by conduction is expressed as

$$q_{cond} = k_d A (T_d - T_o) \tag{4.11}$$

The heat dissipation by convection is given by

$$q_{con} = h_r A_s (T_s - T_o) \tag{4.12}$$

The convective heat transfer coefficient is expressed as

$$h_r = 0.1 \left(\frac{k_a}{D} \right) R_e^{\frac{2}{3}} \tag{4.13}$$

The Reynolds number depends on the velocity of the moving vehicle. The Reynolds number is expressed as

$$R_E = \frac{\rho_a D V}{\mu_a} \tag{4.14}$$

The heat dissipation to the surrounding by radiation is expressed as

$$q_{rad} = A_s \sigma . \varepsilon \left(T_s^4 - T_\alpha^4 \right) \tag{4.15}$$

4.2 Experimental Procedures

4.2.1 Fabrication of Brake Drum Test Rig

A brake drum test rig has been fabricated for the test as shown in Fig. 4.1. The test rig has the provision to mount the brake drum and to rotate it at different speeds. The test rig has been designed by considering the maximum mass of the vehicle. A controller is included in the motor circuit to adjust the load and speed of the motor. A commercial carrier plate which contains the shoe assembly has been mounted on the drum shaft using a bearing and it is kept stable using a spring balance. A master cylinder with a pressure gauge has been mounted on the test rig to apply different braking pressures on the rotating drum. When the brake force is applied, the brake shoes expand and the shoes are forced against the rotating brake drum producing a twisting moment at the carrier plate. This develops a brake torque at the friction surface which has been measured using a spring balance. The applied

Fig. 4.1 Brake drum test rig

brake force has been measured from the pressure gauge. Thermocouples are inserted in the drum as per the SAE specifications to measure the temperature rise during braking.

4.3 Results and Discussions

4.3.1 Dynamic Friction Coefficient

Tests have been conducted to determine the dynamic friction coefficient, brake torque and the temperature rise. During braking, variation of contact has been observed due to applied braking force, sliding speed and temperature at the interface (Eriksson et al. 2002). This variation significantly affects the friction coefficient between the drum and the lining material. The friction coefficient of the new drum and the lining is less but increases with time of operation. This is due to increase in contact area resulting from smoothening of the drum surface and wear of the contact plateaus (Ostermeyer 2003). Dynamic tests have been conducted using a constant brake line pressure (10 bar), braking time of 2 min. The same test is repeated and the corresponding average friction coefficient has been found. The ratio between the applied force and the drag force developed in the brake drum is constant and is defined as the brake factor. The total brake factor of a leading—trailing shoe is expressed as (Limpert 1999).

$$BF = \frac{2\mu\frac{h}{b}}{1 - \left[\frac{\mu.c}{b}\right]^2} \tag{4.16}$$

From the dimensions of the brake drum, the brake factor is computed for different values of friction coefficient. The variation of brake factor with the dynamic friction coefficient is shown in Fig. 4.2. The variation of brake torque with dynamic friction coefficient at different brake line pressures is shown in Fig. 4.3. The dynamic friction coefficient of any brake drum and lining combination can be determined by measuring the torque developed for a certain value of applied pressure as shown in Fig. 4.4. The brake drum is rotated at a speed of

Fig. 4.2 Variation of brake factor with friction coefficient

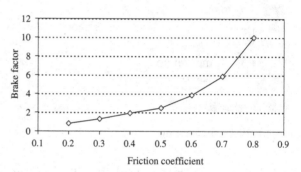

Fig. 4.3 Variation of brake torque with friction coefficient

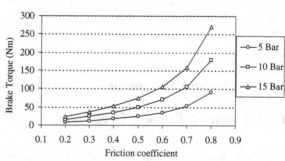

Fig. 4.4 Determination of dynamic friction coefficient

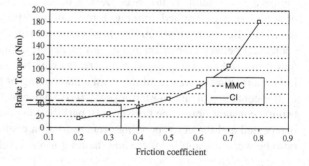

500 rpm and a pressure of 10 bar is applied to the wheel cylinders using the master cylinder arrangement. The corresponding brake torques for cast iron and the MMC brake drums have been observed as 38 and 47 Nm respectively. Horizontal lines are drawn from these values of brake torque to meet the curves drawn at 10 bar. From these meeting points, vertical lines are drawn and those values directly give the dynamic friction coefficient. The value of dynamic friction coefficient of MMC brake drum has been found as 0.475. Similarly, the dynamic friction coefficient of the cast iron brake drum has been observed as 0.425.

4.3.1.1 Determination of Brake Torque

The variation of brake torque with brake line pressures has been also measured. Tests have been conducted for a brake drum speed of 500 rpm which corresponds to a vehicle speed of 50 km/h. The brake torque has been determined for cast iron and MMC brake drum. The brake torque absorbed in MMC brake drum has been observed to be more than the cast iron drum. This is because of the high value of dynamic friction coefficient of MMC which improves the braking performance. This will help in reducing the brake force required for the braking.

4.3.1.2 Validation of Analytical Results

The input parameters for the analytical model are shown in Table 4.1 The variation of average brake torque with brake line pressure for cast iron and the MMC brake drum are shown in Fig. 4.5a, b respectively. In these two cases, little

Table 4.1 Input parameters for the brake drums

Input parameters	Brake drum	
	Cast iron	Al MMC
Inner diameter of brake drum (m)	0.18	0.18
Width of brake shoe (m)	0.03	0.03
Contact angle per shoe	113	113
Brake drum dimension 'h' (m)	0.15	0.15
Brake drum dimension 'b' (m)	0.075	0.075
Brake drum dimension 'c' (m)	0.077	0.077
Diameter of wheel cylinder (m)	0.016	0.016
Wheel cylinder efficiency	0.96	0.96
Density of brake drum (kg/m^3)	7,228	2,785
Density of shoe lining (kg/m^3)	2,034	2,034
Thermal conductivity of brake drum (Nm/h km)	1,74,465	3,74,400
Thermal conductivity of lining (Nm/h km)	4,174	4,174
Specific heat of brake drum (Nm/kg k)	419	970
Specific heat of brake shoe lining (Nm/kg k)	1,256	1,256

Fig. 4.5 Variation of brake torque with *brake line* pressure **a** Cast iron brake drum **b** MMC brake drum

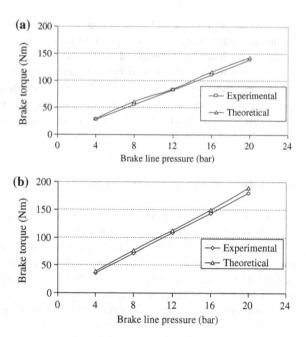

deviation has been observed between the analytical and the experimental values. It has also been observed that the MMC brake drum absorbs more torque than the cast iron brake drum for the same applied brake line pressure. Thus, the MMC brake drum is capable of providing enhanced braking performance than the cast iron brake drum. This is because of the high value of friction coefficient which improves the braking performance. This will also help in reducing the brake force requirements for the braking. The temperature rise during braking has been observed from the temperature indicator and the thermocouple arrangement. The variation of temperature rise with different braking time for the cast iron and the MMC brake drum is shown in Fig. 4.6a, b respectively. Good agreement has been observed between the theoretical and experimental results. The temperature rise in the MMC brake drum is observed as 5–10 °C more than the cast iron brake drum due to the high value of dynamic friction coefficient of the MMC brake drum.

4.3.1.3 Temperature Rise in Brake Drum During Braking

The brake system is subjected to single stop braking in case of emergency stop. During this process, the braking time is less than the time required for the heat to penetrate through the brake drum. So, the heat is completely absorbed by the brake drum and the brake shoe lining. Due to traffic density in the urban perimeter, the vehicle is also subjected to repeated deceleration and acceleration. In this type of braking, alternative heating and cooling of the brake drum occurs. Also, when a vehicle descents from a mountain, the brake is continuously applied over a long

Fig. 4.6 Variation of
temperature rise with braking
time **a** cast iron brake drum
b MMC brake drum

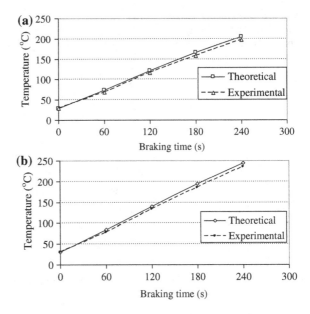

period. Under this condition, the heating and cooling of the brake drum occurs simultaneously. The temperature rise for each braking condition is presented in the following sections.

4.3.1.4 Single Stop Braking

In a single stop or emergency braking, the braking time is less than the time required for the heat to penetrate through the brake drum. Under this circumstance, no convective cooling occurs and all the heat is assumed to be absorbed by the lining and the brake drum. The variation of surface temperature with braking time for the cast iron and the MMC brake drum is shown in Fig. 4.7a, b respectively. The surface temperature is found to increase with braking time and the vehicle speed. For MMC brake drum, the temperature rise is observed to be lower at all the braking time and vehicle velocities. The variation of stopping distance with brake line pressure for the cast iron and the MMC brake drum is shown in Fig. 4.8a, b respectively. For the same velocity, the stopping distance is reduced by increasing the applied pressure without locking of the brake drum. For a velocity of 90 km/h, when a brake line pressure of 50 bar is applied, the stopping distance for cast iron drum has been observed as 64 m. For the same condition, the stopping distance for MMC brake drum has been observed as 51 m.

The variation of stopping time with brake line pressure for the cast iron and the MMC brake drum is shown in Fig. 4.9a, b respectively. For the MMC brake drum, the stopping time has been observed to be lower than the cast iron brake drum under identical braking conditions. The variation of temperature with velocity of

Fig. 4.7 Variation of surface temperature with braking time **a** Cast iron brake drum **b** MMC brake drum

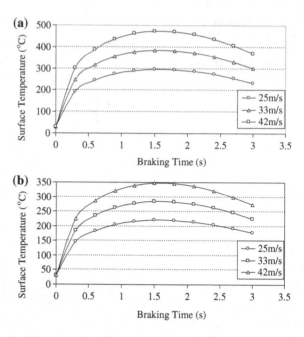

Fig. 4.8 Variation of stopping distance with *brake line* pressure **a** Cast iron brake drum **b** MMC brake drum

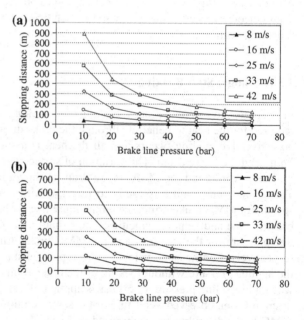

vehicle and friction force for cast iron and the MMC brake drum is shown in Fig. 4.10a, b respectively. The temperature response is found to be more for friction force than for the vehicle velocity. Under identical braking conditions, the

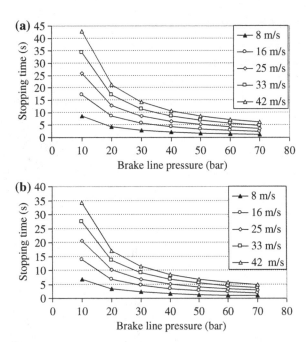

Fig. 4.9 Variation of stopping time with *brake line* pressure **a** cast iron brake drum **b** MMC brake drum

temperature rise has been observed to be low for MMC brake drum while braking under identical conditions. The deceleration of the vehicle with applied brake line pressure for the cast iron and the MMC brake drum is shown in Fig. 4.11. At all the brake line pressures, the deceleration is observed to be higher for the MMC brake drum. The variation of temperature rise at the brake drum surface with vehicle velocity for cast iron and MMC brake drum is shown in Fig. 4.12. For single stop braking at all speeds, the temperature rise at the brake drum surface has been observed to be lower for MMC brake drum. The variation of temperature rise in the brake drum with braking time has been computed for cast iron and MMC brake drum. For all braking times, the temperature rise in MMC brake drum is observed to be lower than the cast iron brake drum as shown in Fig. 4.13. From all the above discussions, it has been observed that the temperature rise in MMC is lower than the cast iron brake drum.

4.3.2 Continuous Braking

When the brakes are applied on a vehicle during downhill descend, the brakes have to absorb potential and kinetic energy continuously. The generation and dissipation of heat takes place simultaneously. The variation of temperature rise with braking time in the cast iron and the MMC brake drum is shown in Fig. 4.14a, b

Fig. 4.10 Variation of
temperature rise with speed
and friction force **a** Cast iron
brake drum **b** MMC brake
drum

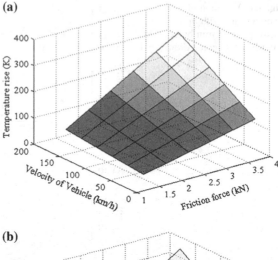

(a)

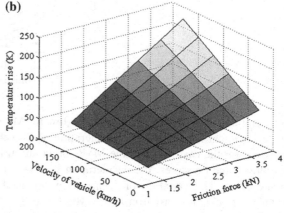

(b)

Fig. 4.11 Variation of
deceleration with *brake line*
pressure

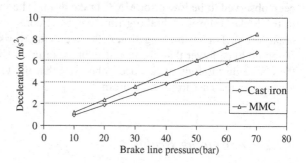

respectively. During continuous braking, the temperature rise in MMC brake drum
is observed to be 2–6 % more than the cast iron brake drum. The increase of
temperature is due to high brake power absorbed by the MMC brake drum.

Fig. 4.12 Variation of temperature rise with speed

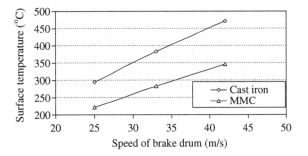

Fig. 4.13 Comparison of temperature rise in brake drums

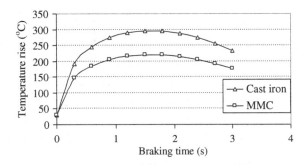

Fig. 4.14 Variation of temperature rise with time during continuous braking
a Cast iron brake drum
b MMC brake drum

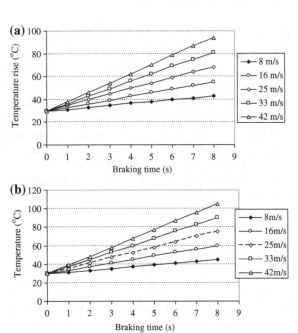

Fig. 4.15 Variation of
temperature rise with time
during repeated braking
a cast iron brake drum
b MMC brake drum

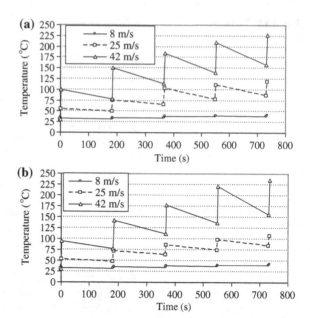

4.3.3 Repeated Braking

Increased vehicle population in cities envisages the need for frequent repeated
deceleration and acceleration of vehicles. During repeated braking, the vehicle is
decelerated until the velocity reduces to zero and is again accelerated to the initial
speed and the next braking cycle is repeated. During this braking condition, the
heat is generated and dissipated repeatedly. The variation of temperature rise in the
cast iron and the MMC brake drum for this condition is shown in Fig. 4.15a, b
respectively. The temperature has been observed to increase during braking and
reduce when the brakes are released. The temperature rise in the MMC brake drum
is found to be 3–7 % higher than the cast iron drum while braking under identical
conditions. It is due to the lower mass and high friction coefficient of the MMC
brake drum.

4.4 Summary

The fabrication of experimental set-up, testing and thermal analysis of Al MMC
have been presented in this chapter. From the aforementioned analysis the following conclusions can be made. The analytical models have been verified using
experimental results.

An experimental setup has been fabricated to test the performance of the cast
iron and the Al MMC brake drums.

The dynamic friction coefficient of cast iron and the MMC brake drum are found to be 0.425 and 0.475 respectively.

When tested at a brake line pressure of 10 bar for a braking time of 60 s, the temperature rise in MMC brake drum is 10 °C higher than the cast iron brake drum. This is due to higher brake power absorbed in MMC brake drum because of 0.05 increase in dynamic friction coefficient.

During repeated braking, the temperature rise in MMC brake drum is found to be 3–7 % more than the cast iron brake drum.

It is also observed that the temperature rise in MMC brake drum is 2–6 % higher than the cast iron brake drum during continuous braking.

Analytical models have been verified using experimental results. Good agreement has been observed between the analytical and the experimental results.

References

Eriksson M, Bergman F, Jacobson S (2002) On the nature of tribological contact in automotive brakes. Wear 252:26–36

Guner R, Yavuz N, Kopmaz O, Ozturk F, Korkmaz I (2004) Validation of analytical model of vehicle brake system. Int J Veh Des 35(4):340–348

Ilinca A, Ilinca F, Falah B (2001) Numerical and analytical investigation of temperature distribution in a brake drum with simulated defects. Int J Veh Des 26(2/3):146–160

Limpert R (1999) Brake design and safety. 2nd edn. SEA, Pennsylvania

Ramachandra Rao VTVS, Rajaram LS, Seetharamu KN (1993) Temperature and torque determination in brake drums. Sadhana 18:963–983 Part 6

Roberts SG, Day TD (2000) Integrating design and virtual test environments for brake component design and material selection. SAE technical paper series 2000-01-1294, Warrendale, PA 15096-0001, USA

Ostermeyer GP (2003) On the dynamics of the friction coefficient. Wear 254:852–858

Chapter 5
Wear Analysis

5.1 Introduction

The wear and friction measurements have been proposed by various researchers (Anderson et al. 1984). Deuis et al. (1997) have presented a review on the dry sliding wear of aluminium composites. The friction and wear mechanisms of aluminium composites and the influence of applied load, sliding speed, wearing surface hardness, model for wear volume and the role of reinforcement phase are also have been presented. Tjong et al. (1997) have conducted wear tests on compocast aluminum silicon alloys reinforced with low volume fraction of SiC. Based on the wear test conducted on the block on ring, they have concluded that the addition low volume fraction of SiC particles is an effective way of increasing the wear resistance of the matrix alloy. Kwok and Lim (1999) have investigated the friction and wear behavior of four Al/SiC$_p$ composites over a wide range of sliding conditions by the use of a specially adapted high speed tester of the pin on disc configuration. In this chapter, the wear behaviour of Al MMC sliding against brake shoe lining material has been observed and compared with the conventional grey cast iron. The wear tests have been carried out on a pin on disc machine, using pin as brake shoe lining material and discs as A356/25SiC$_p$ Al MMC and grey cast iron materials. Pins of 10 mm diameter have been machined from a brake shoe lining of a commercial passenger car. The grey cast iron disc has been machined from a brake drum of a commercial passenger car. The Al MMC disc has been manufactured by dispersion casting technique and machined to the required size. The friction and the wear behaviour of Al MMC, grey cast iron and the brake shoe lining have been investigated at different sliding velocities, loads and sliding distances. The worn of the MMC, the cast iron and the lining have been observed using optical micrographs. The present investigation shows that the Al MMC have considerable higher wear resistance than conventional grey cast iron while sliding against automobile friction material under identical conditions. The wear grooves

© The Author(s) 2015
N. Natarajan et al., *Metal Matrix Composites*, SpringerBriefs in Manufacturing
and Surface Engineering, DOI 10.1007/978-3-319-02985-6_5

formed on the lining material while sliding against MMC and the cast iron have been observed using optical micrographs.

5.2 Materials for Wear Test

The wear behaviour of brake drum material and its counterface friction material are determined in this investigation. Asbestos based brake shoe lining material and two brake drum materials are considered for the test. One is the commercially used gray cast iron and the other is the proposed $A356/25SiC_P$.

5.2.1 Material for Wear Disc

The cast iron and the MMC disc used for the test are shown in Figs. 5.1 and 5.2 respectively. The cast iron disc is machined from a commercial passenger car brake rotor. The inner diameter, the outer diameter and the thickness are 180 mm, 140 mm and 4 mm respectively. The composition is shown in Table 5.1.

The MMC is manufactured through the dispersion casting process. The microstructure of the MMC is shown in Fig. 5.11d. The casting is then finished to a size of outer diameter, inner diameter and thickness as 180, 110 and 5 mm respectively. The composition of the aluminium alloy is shown in Table 5.2.

Fig. 5.1 Cast iron disc

Fig. 5.2 MMC disc

Table 5.1 Composition of grey cast iron

Constituent	Fe	C	Mn	P	S	Si
Percentage	93	3.2–3.5	0.6–0.9	0.12	0.15	2.2

Table 5.2 Composition of A356 aluminium alloy

Constituent	Al	Si	Mg	Cu	Mn	Fe	Ti	Zn
Percentage	90–93	6–7.5	0.45	0.25	0.35	0.6	0.25	0.35

Table 5.3 Composition of automobile friction material

Constituent	Phenolic resin	Asbestos fiber	Cu	Zn	Fe	Others
Percentage	30	45	4	3	4	14

Fig. 5.3 Pin

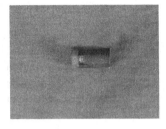

5.2.2 Material for Pin

To study the effect of the candidate Al MMC material on the counter-face friction material, a commercial semi-metallic brake shoe lining material of a passenger car is used as the pin for the wear test. The pin is machined and mounted on a 10 mm diameter rod as shown in Fig. 5.3 for mounting it on the machine. The surface is polished by using A320 emery paper. The surface is cleaned and conditioned before the starting of every experiment. The composition of the lining material is shown in Table 5.3.

5.3 Wear and Friction Coefficient

The material pair used for the brake drum applications should have higher and stable friction coefficient and superior wear resistance.

5.3.1 Frictional Force

For applications like brake drum, the MMCs should withstand high braking forces without undue distortion, deformation or fracture during braking and should maintain controlled friction and wear over long periods. During braking, the temperature rises significantly because of the tribological interactions between the brake drum and the lining. In MMCs, the dispersed particles are helpful in retaining the high temperature strength of the matrix. The frictional force is due to the force of adhesion and deformation. The adhesion is because of Van der Waals forces, dipole interactions, hydrogen bonding and electric charges. The force of deformation is due to polymer asperities, loss of energy due to hysteresis and grooving by the counterface. The friction coefficient depends on material properties like hardness, yield strength, microstructure and surface finish. Liu (1980) has investigated that the frictional force during sliding, is to be the power function of applied load and sliding velocity at a particular temperature as

$$F = \mu(T)P^{a(T)}V^{b(T)} \tag{5.1}$$

where F is the frictional force in Newtons, $\mu(T)$ is the friction coefficient at temperature T, P is the applied load in Newtons, V is the sliding velocity in metres per second, $a(T)$ is the load factor at temperature (T) and $b(T)$ is the velocity factor at temperature (T). Under heavy braking conditions, the value of frictional force reduces due to rise in temperature and result in brake fade.

5.3.2 Wear Coefficient of Disc Materials

Grey cast iron is the workhorse of brake drum applications in automobiles. The tribo system between cast iron and lining material is very complex. The investigation of wear behaviour of these materials while sliding against the brake shoe lining material is timely needed before using it in actual applications. Liu (1980) in his wear theory, has proposed the wear volume as a function of normal load, sliding velocity and hardness of the material as

$$V = \frac{KPL}{3H} \tag{5.2}$$

where V is the volume of material worn in m^3, K is the wear coefficient (to be experimentally determined), P is the applied load in Newton, L is the sliding distance in metre and H is the hardness of the material. But in most investigations, the proposed proportionality between wear volume and load is not always observed.

5.3.3 Wear Coefficient of Lining Material

Friction materials are composites of polymers containing reinforcements, fillers and binders. The reinforcements are metal, glass, acrylic and asbestos fibers. Barite and Aramid are the filler materials. Phenolic resins are used as binders. Because of the composite nature, the tribological behaviour is very complex. Rhee (1980) has derived an equation for the wear of lining material. He has investigated that the wear volume is proportional to the power functions load, sliding velocity and sliding time,

$$V = K_f P^a V^b t^c \tag{5.3}$$

where V is the wear volume in m^3, K_f is the wear coefficient of friction material (to be experimentally determined), P is the applied load in Newton, V is the sliding velocity in m/s, t is the time of sliding in seconds and a, b and c are exponents that depend on material and sliding conditions.

5.4 Experimental Procedures

The wear tests have been conducted on a Ducom pin on disc machine shown in Fig. 5.4. The cast iron and the proposed Al MMC disc have been mounted on the machine. The lining material in the form of pin has been fixed on a holder, which has a provision for applying the load. A balance having an accuracy of 0.1 mg with a maximum weighing capacity of 200 g is used to determine the mass of cast iron disc, MMC disc and the lining material (pin). The machine is connected to a controller and a computer to control and measure sliding velocity, sliding time and frictional force. The disc and the lining material have been weighed before and after each test and the weight loss has been used as the measure of wear. The frictional forces are recorded in the computer. Although the frictional force varies with sliding time, an average value is considered for the analysis. The wear test is conducted by varying the load and keeping the speed and sliding distance as constant. The above procedure is repeated for different speeds and sliding

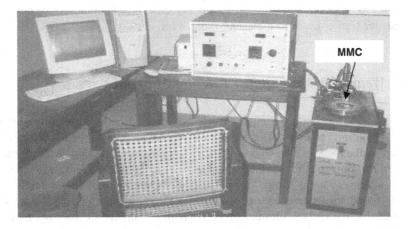

Fig. 5.4 Experimental set-up used for wear test

distances. To investigate the wear mechanisms and characteristics of transfer layer, the worn debris and the wear tracks have been analysed using optical micrographs. The damaged surfaces of disc and lining material have been analysed by the optical micrographs.

5.5 Results and Discussions

5.5.1 Wear of Cast Iron Sliding Against Friction Material

In the first phase, the wear of cast iron and the friction lining pin material have been determined from several tests conducted at different loads and speeds. In most of the investigations, the wear is expressed in mass loss. In the present investigation, since the materials have three different densities, the expression of wear in terms of mass will not be useful for comparative purpose. The wear in terms of volume loss will be useful in order to determine the geometrical changes in the components. So, the wear loss is expressed in terms of volume loss in this present investigation. The wear loss is measured by changing the applied load on the lining pin by keeping the sliding speed and the sliding distance as constants. The variation of wear with load for cast iron disc while sliding against friction lining is shown in Fig. 5.5a. The wear is low at lower value of applied loads and increases with load at constant a ratio according to Archard equation of wear (5.1). The wear is due to the nature of contact of the sliding couple. At lower loads, the contact plateaus and temperature rise are low. As the applied load is increased, the wear loss is found to increase. Higher wear is observed for the maximum load. The wear is found to increase with sliding velocity. The same trend is also observed for the increased sliding velocity and is shown in Fig. 5.6a. As the sliding velocity is increased the transfer film is destroyed at faster rate and new film is to be formed to compensate for this, thereby enhancing the wear. The higher contact temperature developed during high load and sliding velocity at the friction surface destroys the transfer film at faster rate causing more wear.

5.5.2 Wear of MMC Against Friction Material

The wear of MMC sliding against the friction material is determined for various loads and sliding velocities. The variation of wear with applied load is determined by keeping the load and the sliding velocity as constants. The same experiment is repeated for different sliding velocities. The wear is found to increase with applied load at a slower rate as shown in Fig. 5.5b. For increase of sliding velocity, the wear is found to increase and it is shown in Fig. 5.6b. In case of MMCs, it is observed that the surface film is formed on both the sliding surfaces but more at the MMC surface.

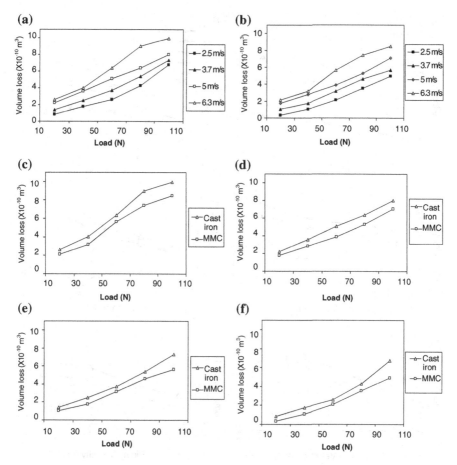

Fig. 5.5 Variation of wear in cast iron and MMC with applied load. **a** Cast iron. **b** MMC. **c** Wear at 6.3 m/s. **d** Wear at 5 m/s. **e** Wear at 3.7 m/s. **f** Wear at 2.5 m/s

5.5.3 *Wear Comparison of Cast Iron and MMC*

The comparison of wear of cast iron and the MMC sliding against the friction material under identical conditions are shown in Fig. 5.5c–f. In all these cases, the wear is found to increase with applied load and speed. But the wear is observed as 1.5 times more for cast iron. For MMCs, the wear is found to be low, because of the presence of the hard SiC particles which act as the load bearing member and abrasive in nature. The variation of wear with sliding velocity for the cast iron and the MMC are shown in Fig. 5.6c–f. In all these comparisons, the wear is observed to be more for the cast iron material.

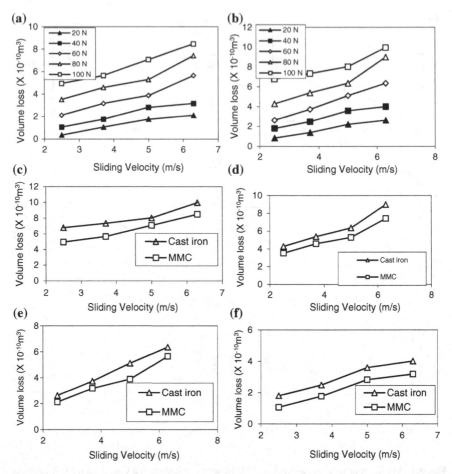

Fig. 5.6 Variation of wear in cast iron and MMC with sliding velocity. **a** Cast iron. **b** MMC. **c** Wear at 100 N. **d** Wear at 80 N. **e** Wear at 60 N. **f** Wear at 40 N

5.5.4 Wear of Friction Material Against Cast Iron

The wear of friction material is measured before and after every test and the results are analysed. From Fig. 5.7a, it is observed that the wear of lining material is found to increase with applied load and this increase is high for higher loads. At higher load, the friction material is forced against the disc resulting in high temperature at the interface, thereby destroying the transfer film at a faster rate. So, new transfer films are formed at faster rate enhancing the wear of the lining. The variation of wear with sliding velocity is shown in Fig. 5.8a.

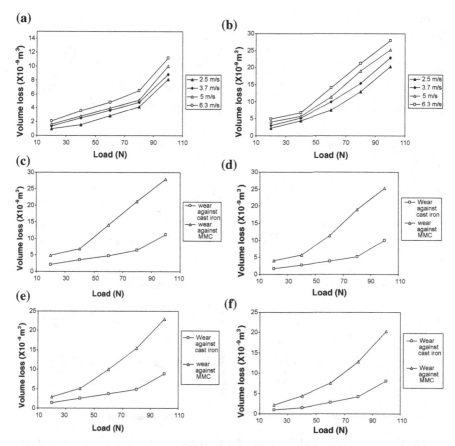

Fig. 5.7 Variation of wear in friction material with applied load. **a** Wear against cast iron.
b Wear against MMC **c** Wear at 6.3 m/s. **d** Wear at 5 m/s, **e** Wear at 3.7 m/s, **f** Wear at 2.5 m/s

5.5.5 *Wear of Friction Material Against MMC*

The variation of wear in lining material with applied load is shown in Fig. 5.7b.
The wear is observed observed to increase with applied load and is very high for
higher loads. The higher wear rate is due to the presence of the silicon carbide
particles in the counterface material. The protruding silicon carbide particles from
the counterface destroy the transfer film and plough the lining material. The
variation of wear with sliding velocity is shown in Fig. 5.8b. The wear is more
influenced by the sliding velocity, because the ploughing is fast at higher sliding
velocities.

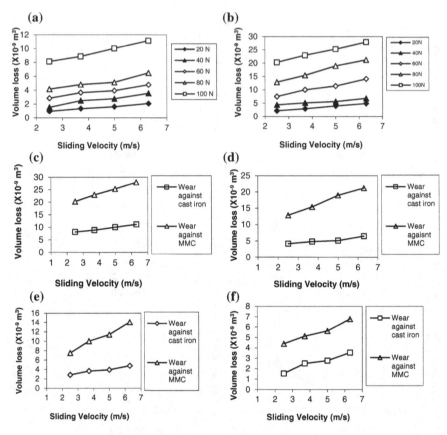

Fig. 5.8 Variation of wear in friction material with sliding velocity. **a** Sliding against cast iron.
b Sliding against MMC. **c** Wear at 100 N. **d** Wear at 80 N, **e** Wear at 60 N. **f** Wear at 40 N

5.5.6 Comparison of Wear in Friction Material

The comparisons of wear in the friction material sliding against cast iron and the
MMC under identical conditions are presented in Fig. 5.7c–f. From results of all
the figures, it is observed that the wear of friction material is found to increase with
applied load in both the cases, but the wear is observed to be more for the friction
material sliding against the MMC. This trend is observed in all cases. The comparisons of wear at different sliding velocities are shown in Fig. 5.8c–f. It is
observed that the variation of wear is less for lining material with sliding velocities. But in all the cases, it is observed that the wear is more for the friction lining
while sliding against the Al MMC counter part.

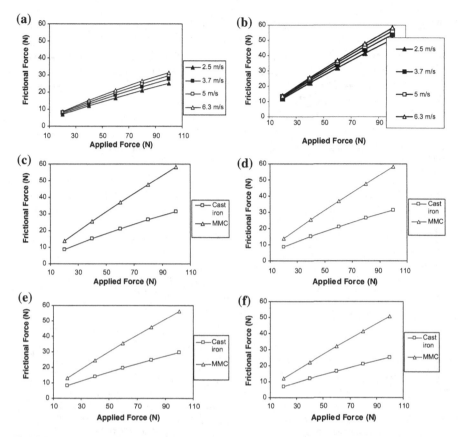

Fig. 5.9 Variation of frictional force with applied force. **a** Sliding against cast iron. **b** Sliding against MMC, **c** Frictional force at 6.3 m/s. **d** Frictional force at 5 m/s. **e** Frictional force at 3.73 m/s. **f** Frictional force at 2.5 m/s

5.5.7 Frictional Force

The variation of frictional force with applied load for cast iron and friction material couple is shown in Fig. 5.9a. More variations are observed with applied load than with the sliding velocity. At higher loads, the frictional force is higher because of more contact area at the friction material surface. The variation of frictional force while sliding against the MMC is shown in Fig. 5.9b. Here, higher variations are observed for loads and lower variations for the variation of sliding velocities. The comparison of frictional force developed by these materials under identical conditions is presented in Fig. 5.9c–f. In all these cases, the frictional force is observed as high for the MMC and the friction material sliding couple.

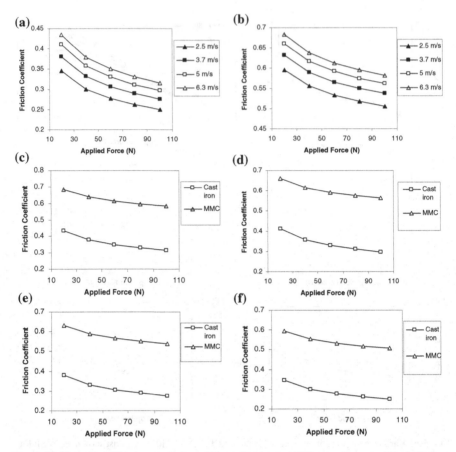

Fig. 5.10 Variation of friction coefficient with applied load. **a** Sliding against cast iron. **b** Sliding against MMC. **c** Friction coefficient at 6.3 m/s. **d** Friction coefficient at 5 m/s. **e** Friction coefficient at 3.73 m/s. **f** Friction coefficient at 2.5 m

5.5.8 Friction Coefficient

There exists a definite ratio between the force developed and the applied force, called the friction coefficient. A stable and higher friction coefficient is essential for brake drum applications. The variation of friction coefficient for cast iron and the lining material couple is shown in Fig. 5.10a. The friction coefficient is observed as high for lower loads and reduced for increase of applied loads. This is because at lower loads, the transfer film is found to be stable for more time and temperature rise is also low, whereas at higher loads the transfer film is destroyed at faster rate and the temperature rise is also high. The variation of friction coefficient for MMC and lining couple is shown in Fig. 5.10b. Similar variations are observed for the variations of applied load as mentioned above. The

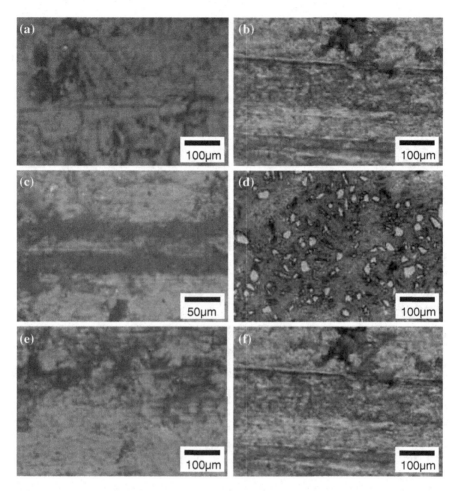

Fig. 5.11 Optical micrographs. **a** Cast iron surface before wear. **b** Cast iron surface after wear. **c** Cast iron surface after wear. **d** Microstructure of MMC. **e** MMC surface before wear. **f** MMC surface after wear

comparison of friction coefficient for cast iron and the MMC is shown in Fig. 5.10c–f. In all these observations, the friction coefficient is observed to be 1.25 times more for the Al MMC while sliding under identical conditions.

5.6 Optical Micrograph of Contact Surfaces

The optical micrographs of the contact surfaces of cast iron, the MMC and the lining material are analysed before and after the wear test. The contact surface of cast iron before wear test is shown in Fig. 5.11a. The optical micrograph of the

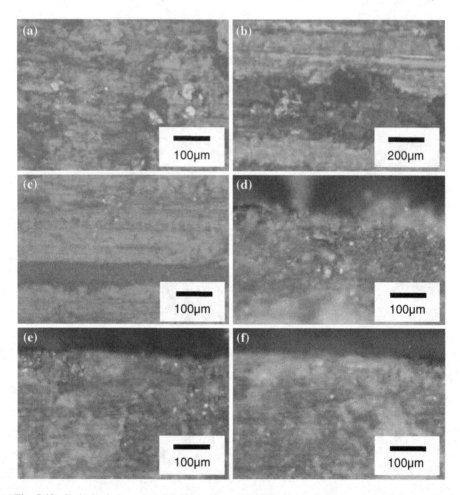

Fig. 5.12 Optical micrographs of lining surfaces. **a** Lining surface before wear. **b** Lining surface after wear against CI. **c** Lining surface after wear against MMC. **d** Cross section showing top surface before wear test. **e** Cross section showing the top surface after wear against CI. **f** Lining surface after wear against MMC

contact surfaces after sliding over a distance of 2,000 m for an applied load of 60 N and a sliding velocity of 5 m/s is shown in Fig. 5.11b, c. The worn surfaces show the wear traces formed on the sliding surface as shown in Fig. 5.11c. The optical micrographs of the MMC before and after wear test are shown in Fig. 5.11e, f respectively. The optical micrographs of lining material before and after wear against cast iron are shown in Fig. 5.12a, b respectively. The microstructure showed in Fig. 5.12a reveals the composite nature of lining material. Figure 5.12b shows the wear.

Traces formed on the lining material while sliding against cast iron. It shows the small wear grooves and transfer films formed on the surface. The worn surfaces while sliding against MMC material is shown in Fig. 5.12c. From the microstructure it is observed that the grooves are more in width and depth than the surface of the lining material sliding against cast iron. The formation of wear grooves along the lining surface enhances the wear of lining, hence the wear is observed to be high. To study the nature of top surface, the cross section is taken in a direction perpendicular to the sliding direction. The cross section of lining before wear test is shown in Fig. 5.12d. It shows the irregular top surface showing the contact plateaus. The optical micrographs of cross sections after wear against cast iron and MMC is shown in Fig. 5.12e, f respectively. These structures show the thin surface film formed on the sliding surfaces.

References

Anderson AE (1984) Friction and wear of automotive brakes, ASME Hand Book, Wear. 18:569–577

Deuis RL, Subramanian C, Yelllup JM (1997) Dry sliding wear of aluminium composites–a review. Compos Sci Technol 57:415–435

Kwok JKM, Lim SC (1999) High speed tribological properties of some Al/SiC composites: I frictional and wear rate characteristics. Compos Sci Technol 59:55–63

Liu T, Rhee SK, Lawson KL (1980) A study of wear rate and transfer films of friction materials. Wear 60:1-12

Tjong SC, Wu SQ, Liao HC (1997) Wear behaviour of an Al-12 % Si Alloy Reinforced with a low volume fraction of SiC particles. Compos Sci Technol 57:1551–1558

Chapter 6
Machinability of Metal Matrix Composites

Metal Matrix Composites (MMC) owing to their increased specific strength and stiffness are replacing the conventionally used materials especially in the fields of automobile, aerospace and structural engineering. Due to their low cost, ceramic particles reinforced aluminium alloy are most popular among the MMC. The ceramic particles or reinforcements such as SiC and alumina make machining of composite tougher. This difficulty in machining opens a wide area of research in the processing of these materials (Teti 2002).

Quan et al. (1999) described that the merchant's circle equation cannot be used for the composite materials as the yield strength of these materials are completely different and also the assumptions made during merchant circle derivation restricts their application to composite materials. They also related the coefficient of chip deformation with the fraction of reinforcement and the shear angle. Ozben et al. (2008) investigated the effects of the reinforcement fraction on the mechanical properties like hardness and tensile strength of the Al/SiCp composite. Effects of the machining parameters and reinforcement fraction on the tool wear and surface roughness during turning of the Al/SiCp composite using TiN coated carbide tool were examined by them. The machinability of Al/SiCp composite while turning using rhombic tools were reported by Manna and Bhattacharayya (2003). BUE and chip formation were examined using the SEM micrographs to provide an economic machining solution through their work.

Davim (2003) reported the influence of the cutting parameters during turning Al/SiCp on the surface roughness, tool wear and power requirements by employing ANOVA and multiple linear regression techniques. PCD turning tools are used to compare the cutting forces obtained from the conventional turning and ultrasonic vibration turning, there by concluding that the low speed and high depth of cuts reduces cutting forces in ultrasonic vibration turning. Muthukrishnan and Paulo Davim (2009) turned Al/SiCp bars using coarse grade PCD inserts under different cutting conditions. The machining conditions were then optimised for minimized surface roughness by employing the ANOVA and ANN. Seeman et al.

© The Author(s) 2015

N. Natarajan et al., *Metal Matrix Composites*, SpringerBriefs in Manufacturing and Surface Engineering, DOI 10.1007/978-3-319-02985-6_6

(2010) developed a mathematical model for machinability evaluation in turning of Al/SiCpcomposites. They also described the effects of process parameters on the tool flank wear and surface roughness.

Ge et al. (2008) used PCD tools for ultra precision turning of Al/SiCp MMC and revealed the effects of the cutting speed and feed rate on the tool workpiece surface integrity. Karakas et al. (2006) studied the effects of the cutting speed on the tool performance in end milling of B4Cp particles reinforced aluminium MMC. They also compared the wear performance of the uncoated tool and multi-layer coated tools and found that the triple layer coated tool has better wear resistance. Oktem et al. (2005) applied response surface methodology (RSM) for optimization of machining parameters for minimized surface roughness during milling of the Al–7075 alloy parts. They also used genetic algorithm (GA) for optimizing the parameters for desired surface roughness. Lin et al. (1998) studied the chip formation in turning of the Al/SiCp MMC using PCD inserts. They analysed the SEM micrographs of the chips formed and found the separation of the matrix and reinforcement within the chip. The chip formation mechanism in turning of the Al/SiCp MMC and Inconel was analysed by various researchers. They tried to correlate the quality of the machined surface by analysing the SEM micrographs of the chips formed during turning. Arokiadass et al. (2011) made an attempt to develop a predictive model of surface roughness in end milling of Al/SiCp MMC.

The present work reports the effect of machining parameters on cutting force, surface finish and behaviour of chip during end milling using carbide inserts. The output parameters such as cutting force and surface finish are optimised using response surface model and artificial neural network. Further this paper reports the effect of coating on the tool life in finish milling of metal matrix composites.

6.1 Experimental Procedure

6.1.1 Material Synthesizing

The 6061 Al alloy is chosen as the metal matrix of the composite and the reinforcement is SiC powders of size 36 μm (added 15 % by weight). Tables 6.1 and 6.2 presents the composition of matrix and reinforcement used for the metal matrix composites.

The Al/SiCp MMC with the above composition is synthesized by stir casting process. Figure 6.1 shows a schematic sketch of the stir casting setup, describing its parts. Table 6.3 presents the parameters used for synthesising the metal matrix specimen. The micro-hardness of the MMC is found to be as 106 VHN using tested Vickers micro-hardness tester at a load of 100 g.

Table 6.1 Composition of Al-6061 alloy

Al	Mg	Fe	Si	Mn	Others
97.25 %	1.08 %	0.17 %	0.63 %	0.52 %	0.35 %

Table 6.2 Composition of SiC powders

SiC	SiO$_2$	Si	Fe	Al	Free C
99.64 %	0.15 %	0.02 %	0.02 %	0.02 %	0.15 %

Fig. 6.1 Stir casting setup

Table 6.3 Experimental details of stir casting

Furnace	Induction furnace
Pre-heater	Electric furnace
Mass of Al 6061 alloy	900 g
Mass of SiC reinforcement	135 g
Mass of Mg (wetting agent)	15 g
Melting point of Al 6061	700 °C
Pre heat temperature of SiC reinforcement	450 °C
SiC addition temperature	750 °C
Stirrer speed	850 rpm
Time of SiC addition	120 min
Post stirring time	20 min
Mould preparation	Wooden pattern in green sand

Using optical microscope the microstructure of the synthesized Al/SiC$_p$ MMC is examined. The distribution of the SiC particles and the agglomerations is presented in the Fig. 6.2.

Fig. 6.2 Microstructure of
Al/SiCp MMC at 200X

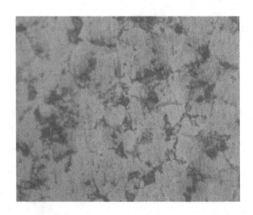

6.1.2 Experimental Setup

The end milling experiments are carried out in a 3 axes CNC Makino Vertical
Machining Centre Model S33. The experimental setup for end milling is presented
in Fig. 6.3. Milling tool dynamometer is mounted on the milling machine table and
the rectangular metal matrix composite specimen is held in its fixture. The
dynamometer is interfaced with personal computer (PC) for force measurements.

Milling is a very complicated cutting process, which involves many parameters
such as cutting speed, feed rate, depth of cut and tool geometry, etc. The most
influential factors affecting the surface finish and forces acting on the tool were
studied by conducting a set of experiments. The factors considered for the experi-
mentation are cutting speed, feed, and depth of cut. The experimental conditions are
presented in Table 6.4. The experiments have been conducted under dry condition in
the vertical machining center. The tool used for end milling is by an inserted cutter of
Ø16 mm made by Sandvik Coromant. The specification of the insert used is R310 11
T3 08E NL-13A. It is an uncoated carbide insert, which is given in Fig. 6.4. One set of
insert is coated using nano-composite structured Hyperlox coating by Physical
Vapour Deposition (PVD) magnetron sputtering technique. The coating thickness is
of 3 μm in order to compare with uncoated insert.

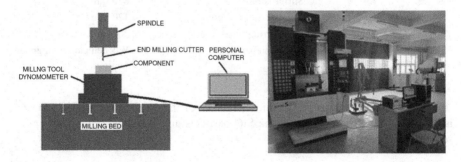

Fig. 6.3 Experimental setup

Table 6.4 Machining parameters

Parameter level	Speed (m/min)	Feed rate (mm/rev)	Depth of cut (mm)
−1	150	0.05	0.10
0	200	0.20	0.15
1	250	0.35	0.20

Fig. 6.4 Details of the insert

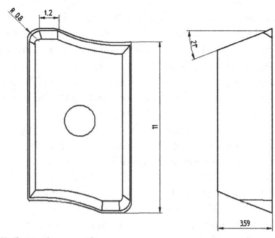

All dimensions are in mm

Using central composite design (CCD), totally 16 experiments are carried out. For each run, the forces acting on the tool are measured using tool dynamometer. The surface roughness is measured by Taylor Hobson surface roughness tester with a cut–off length of 2.5 mm. The surface roughness is measured at 3 different places in a machined surface and average of these values is taken. The machining parameters are optimised using RSM. For the optimised machining condition, the effect of Hyperlox coated inserts on the surface roughness and tool wear are analysed by comparing the results of the experiments conducted using coated and uncoated inserts under dry condition.

6.2 Results and Discussion

Milling can generally be classified as rough milling and finish milling. In case of rough milling the major factor to be considered is the material removal rate whereas in finish milling it is surface roughness. In both the cases the tool life is expected to be longer. The improvement in surface finish and tool life can be accomplished by optimizing the machining conditions or by changing the tool conditions. Here under both the conditions experiments are conducted, the results are discussed below.

6.2.1 Response Surface Regression Analysis

Response surface model (RSM), which is an analytical function, in predicting surface roughness and tool force values is developed using RSM. RSM uses statistical design of experiment (experimental design) technique and least-square fitting method in model generation phase. The initial stage in creating a RS model is to design experiments using one of the RSM designs. Here, Central Composite Design (CCD) is used for experimental design. A face–centred experimental design with 3 factors (speed, feed rate, depth of cut) and 20 runs is generated using CCD in the MINITAB software. Among the 20 runs, 4 centre point runs are ignored and totally 16 runs of experiments are conducted. The tool forces are measured using Syscon Milling tool dynamometer. The forces acting on the tool are shown in the Fig. 6.5. The tangential and feed forces are considered vital over the thrust force. The total force acting on the tool is then calculated as the vector sum of the tangential and feed force.

The surface roughness (Ra) is measured using Taylor Hobson surface roughness tester (shown in Fig. 6.6) with a cut–off length of 2.5 mm. At each slot, the surface roughness is calculated at 3 different spots and their average is taken as the average surface roughness. Table 6.5 shows the list of experiments conducted and the tool force and surface roughness obtained at each run.

The obtained results are analysed using the MINITAB software. The speed, feed rate and depth of cut are chosen as the independent variables or the factors and the surface roughness and tool force are chosen as the dependent variables or

Fig. 6.5 Forces acting on the tool

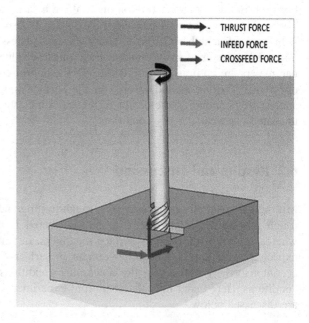

Fig. 6.6 Surface roughness teste

Table 6.5 Experimental results of milling

Trial No.	Cutting speed, V (m/min)	Feed rate, f (mm/rev)	Depth of cut, d (mm)	Total force, F (N)	Surface roughness, Ra (μm)
1	150	0.35	0.10	27.74	1.58
2	150	0.05	0.10	31.02	2.28
3	150	0.05	0.20	13.87	1.33
4	250	0.05	0.20	21.93	1.44
5	200	0.35	0.15	71.41	2.82
6	200	0.20	0.15	59.67	2.25
7	250	0.05	0.10	57.20	1.06
8	250	0.35	0.10	130.88	1.65
9	200	0.20	0.10	136.63	2.04
10	150	0.20	0.15	39.24	2.62
11	200	0.05	0.15	19.62	2.30
12	150	0.35	0.20	49.05	2.65
13	200	0.20	0.15	59.67	2.65
14	250	0.20	0.15	74.71	2.38
15	200	0.20	0.20	69.36	2.34
16	250	0.35	0.20	69.36	2.42

responses. Response surface regression equation for cutting force (F) is given in Eq. 6.1:

$$F = -150.253 + 3.662V + 325.23f - 2762.29d - 0.00704v^2 - 1291.42f^2$$
$$+ 11370.6d^2 + 1.486Vf - 5.046Vd + 203.382fd$$

$$(6.1)$$

Response surface regression equation for surface roughness (Ra) is given in Eq. 6.2:

$$R_a = -1.668 + 0.0144V - 5.303f + 43.254d - 0.0007v^2 - 5.20307f^2$$
$$- 194.828d^2 + 0.0158333Vf + 0.0515Vd + 40.1667fd \qquad (6.2)$$

The correlation coefficient of the regression equations of cutting force and surface roughness are 0.9 and 0.89 respectively.

6.2.2 RS Model Analysis

The developed mathematical models are subjected to ANOVA and F ratio test to justify their goodness of fit. The calculated values of F ratios for lack-of-fit are compared to standard values of F ratios corresponding to their degrees of freedom to find the adequacy of the developed mathematical models. The F ratio is calculated from ratio of mean sum of square of source to mean sum of experimental error. The standard percentage point of F distribution for 95 % confidence level is 4.06 (Seeman 2010). The results of analysis in the Tables 6.6 and 6.7 reveal that the actual F values 1.19 and 1.05 which are less compared to the standard F value and thus adequate within the 95 % confidence limit.

Table 6.6 Analysis of total force RS model

Source	Degrees of freedom	Sum of squares	Adjusted Mean of squares	F value	P value
Regression	9	17275.5	1919.50	6.62	0.016
Linear	3	10482.7	836.69	2.89	0.125
Square	3	4505.9	1501.98	5.18	0.042
Interaction	3	2286.9	762.30	2.63	0.145
Residual error	6	1739.2	289.86	–	–
Lack of fit	5	1489.6	297.91	1.19	0.589
Pure error	1	249.6	249.64	–	–
Total	15	19014.7	–	–	–

Table 6.7 Analysis of surface roughness RS model

Source	Degrees of freedom	Sum of squares	Adjusted mean of squares	F value	P value
Regression	9	3.81293	0.42366	5.10	0.030
Linear	3	1.20891	0.13437	1.62	0.281
Square	3	1.63259	0.54420	6.56	0.025
Interaction	3	0.97144	0.32381	3.90	0.073
Residual error	6	0.49801	0.08300	–	–
Lack of fit	5	0.41801	0.08360	1.05	0.627
Pure error	1	0.08000	0.08000	–	–
Total	15	4.31094	–	–	–

The contribution of the cutting parameters on the dependent Variables (DV) namely surface roughness and tool force can be studied by Pareto chart which gives the significance of each term in the generated RS model in terms of standardized effect estimate. Using STATISTICA 10.0, the Pareto chart is obtained for the total force and surface roughness RS models as shown in Figs. 6.7 and 6.8 respectively.

In the Pareto chart, L denotes that the term is linear, Q denotes that the term is quadratic. Also 1, 2 and 3 stands for speed, feed and depth of cut respectively. From Fig. 6.7 it is understood that the total force is highly influenced by feed rate

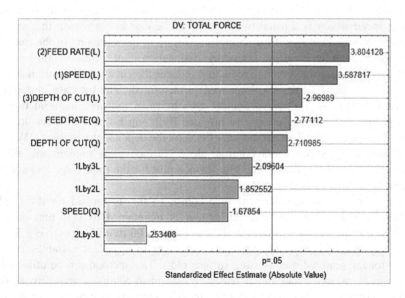

Fig. 6.7 Pareto chart for total force RS model

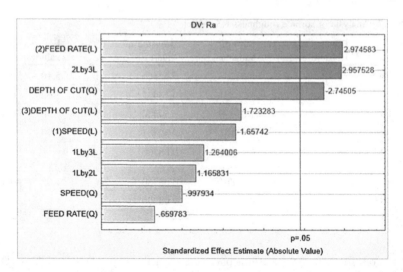

Fig. 6.8 Pareto chart for surface roughness RS model

and speed. This may be due to the fact that the tangential force component opposes the rotational movement of the tool at cutting edges and the feed force is the one that directly opposes the feed motion of the tool. A keen observation of Table 6.5 will reveal the fact that the force changes highly with respect to feed rate compared to other two parameters.

From Fig. 6.8 it is clearly understood that the surface roughness is highly influenced by feed rate and the interacted feed rate and depth of cut. Also from the Table 6.5 it can be observed that the surface roughness decreases with increased cutting speed which is a usual trend. This can be due to the reason that at high cutting speed the heat flow time shortens resulting in the softening of the material and thereby aiding in grain boundary dislocation thus reducing the surface roughness. Whereas the feed rate reduction reduces the cut chip thickness resulting in the lesser cutting forces and improved surface finish. The increase in depth of cut leads to increased abrasive contact area between tool and material thereby causing sufficient heat generation to produce Built -up Edges (BuE) resulting in reduced surface finish. The interaction effects on the dependent variables (DV) can be studied using surface and contour plots. Surface plots explain the effects effectively than the contour plots. The surface plot for the total force and surface roughness are shown in Fig. 6.9.

In plotting the above surface plots, a constant value is applied for third independent variable. In Fig. 6.9a depth of cut is maintained as 0.15 mm, and in Fig. 6.9b, c feed rate is 0.2 mm/rev and cutting speed 200 m/min respectively. So, by fixing one of the independent variables, the other independent variables can be chosen for the required output using surface plots. This method can be utilised for machining at low tool forces and increased surface finish. Similar graphs (Fig. 6.10a–c) were obtained for the effect of machining parameters on surface roughness (Ra).

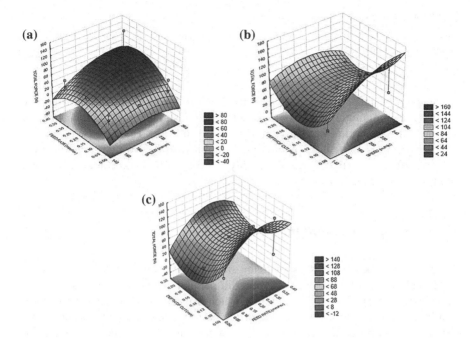

Fig. 6.9 **a** Feed rate, speed (vs.) total force **b** Depth of cut, speed (vs.) Total force **c** Depth of cut, feed rate (vs.) Total force

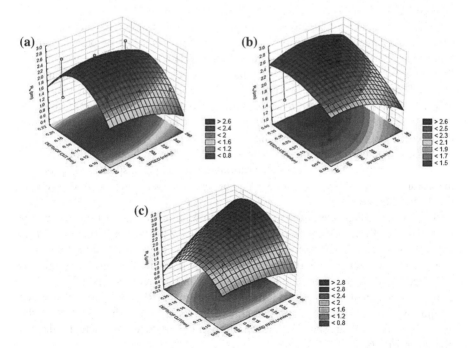

Fig. 6.10 **a** Speed, depth of cut (vs.) R_a **b** Feed rate, speed (vs.) R_a **c** Depth of cut, feed rate (vs.) R_a

Table 6.8 Response surface optimiser result

Speed (m/ min)	Feed rate (mm/ rev)	Depth of cut (mm)	Total force (N)	Surface roughness (μm)
250	0.05	0.20	8.9291	1.2639

Apart from the above analysis optimised values for machining parameters are obtained for minimum tool force and minimum surface roughness using the response surface optimiser. Those values are given in the Table 6.8.

The optimised value has a composite desirability of 0.93 which shows that the multiple–objective optimization is highly reliable. The optimization algorithm used here is "Minimum the Best".

6.2.3 Prediction of Output Using ANN

A neural network is a network of many simple processors (units) each having a small amount of local memory operating in parallel. The units are connected by communication channels (connections), which usually carry numeric data, encoded by one of the various ways. One of the best-known examples of a biological neural network is the human brain. The Artificial Neural Network is developed to try to emulate this biological network for the purpose of learning the solution to a physical problem from a given set of examples. The general architecture of a 3-layered Multi-layer Perceptron (MLP) is shown in Fig. 6.11.

MLP uses back propagation algorithm (BPA) for training the network in a supervised manner. This learning process has two operations or passes namely forward and backward pass. During forward pass, data is read and the weights are fixed at input nodes. During backward pass, the error between the desired and the predicted output updates the weights accordingly. In this way, weight values are adjusted in an iterative fashion while moving along the error surface to arrive at minimal range of error, when input patterns are presented to the network for learning the network (Muthukrishnan 2009).

ANN for the current problem is generated using Neural Network tool (GUI) in MATLAB R2010a. The typical observations of the output response comprises of the parameters as shown in Table 6.9. The network performance is defined by these observations.

The surface roughness results predicted using the generated neural network is shown in the Table 6.10. Comparing these results with the experimental values and values obtained using RS model yields the error in the prediction process.

By using 40 neurons the average error in ANN prediction of surface roughness is reduced to 0.56 %, whereas the average error in RS prediction is 8.55 %. This clearly depicts that ANN offers more accuracy over RS prediction of surface roughness. The maximum error percentage in each case is highlighted in the Table 6.10. A bar chart showing the variation in the predicted and actual values is plotted in Fig. 6.12.

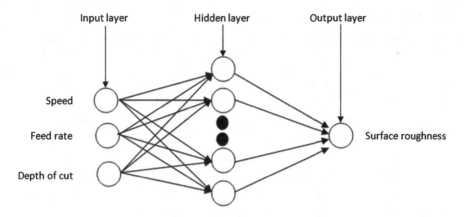

Fig. 6.11 Architecture of neural network

Table 6.9 Typical observations of network performance

Parameter	Description/value
Network configuration	3–40–1
Number of hidden layer	1
Number of hidden neurons	40
Transfer function used	Logsig(sigmoid)
Number of patterns used for training	12
Number of patterns used for testing	2
R value	0.9536
Number of epochs	1,000
Learning factor (η)	0.7
Average error	0.56 %

6.2.4 Chip Formation Mechanisms

Chips can generally be classified as continuous, discontinuous, continuous with BuE and serrated chips. Due to the presence of abrasive particles in the MMC, the chips formed are discontinuous most of the time. But at higher feed rates, saw toothed chips with primary cracks at the outer free surface and secondary cracks at the inner surface with BuE were observed. A few SEM micrographs will aid in visual compliance of the above. The SEM micrograph in Fig. 6.13 shows the top surface of a chip. Cracks are predominantly formed because the 15 % of SiC particle reinforcement has induced high brittleness in the material by reducing its ductility. This results in formation of discontinuous chips.

The chip formation mechanism can be now revealed as initiation of crack from outer free surface due to shear stress induced by tool rake. And during machining at shear zones, the separation of particle and matrix results in voids caused by the

Table 6.10 Comparison of ANN, RS and experimental results

Trial No.	Experimental R_a	ANN prediction of R_a	RS prediction of R_a	% Error in ANN prediction	% Error in RS prediction
1	1.58	1.580127	1.794	0.008	13.54
2	2.28	2.279932	2.092	0.003	8.24
3	1.33	1.349071	1.546	1.434	16.24
4	1.44	1.439931	1.264	0.004	12.22
5	2.82	2.820303	2.755	0.010	2.30
6	2.25	2.300046	2.601	2.224	15.60
7	1.06	1.060023	1.295	0.002	22.19
8	1.65	1.649795	1.472	0.012	10.78
9	2.04	2.039954	1.957	0.002	4.06
10	2.62	2.620057	2.575	0.002	1.71
11	2.30	2.352788	2.213	2.295	3.78
12	2.65	2.650007	2.453	0.0002	7.43
13	2.65	2.650046	2.601	0.001	1.85
14	2.38	2.380136	2.273	0.005	4.49
15	2.34	2.340046	2.271	0.002	2.94
16	2.42	2.490259	2.646	2.903	9.34

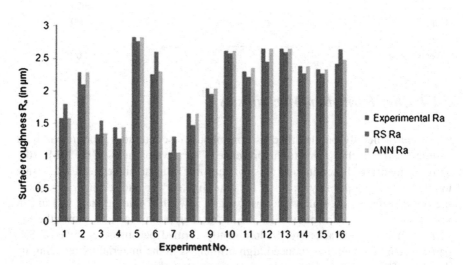

Fig. 6.12 Comparison of ANN and RS prediction

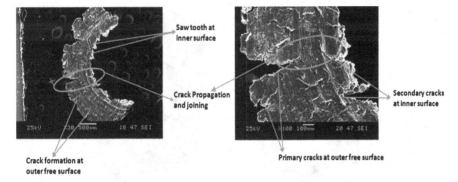

Fig. 6.13 Micrograph of chips showing crack formation

stress concentration at the edge of particles (Lin 1998). These stress components along with the help of voids results in crack propagation and discontinuous chip formation. The saw tooth profile is formed due to highly strained inner surface where the reinforcement is coarsely distributed (Fig. 6.13).

Figure 6.14 shows the views of sheared (bottom) surface of a chip. It shows a completely propagated chip, which is about to break. We can also see that aluminium is pasted to the edges of the chip. This localized melting of aluminium occurs at the shear zones where the reinforcement concentration is high at the time of material removal. Owing to the low heat conductivity of SiC particles, aluminium absorbs most of the heat from the shear zone and gets melted.

6.2.5 Effect of Hyperlox Coating on Tool Wear

Optimization is generally done in machining of MMC due to the fact that compared with other materials, tool wear soars while machining them. Reasons behind are the mechanical and thermal loads acting on the tool. Mechanical loads are

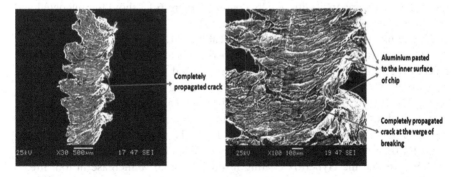

Fig. 6.14 Micrograph of chips showing completely propagated crack

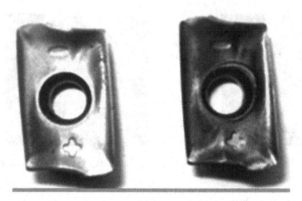

Fig. 6.15 Insert before and after coating

abrasion caused by contact with the reinforcement particles, alternating stress due to inhomogeneity of the material and dynamic loads caused by the reinforcement impacts at the cutting edge. Contributing to the thermal loads are relatively low cutting temperature (limited by the melting point of the Aluminium matrix material when compared with the SiC reinforcement) and high local temperature generated by intensive micro-contact between cutting edge and reinforcement.

Uncoated carbide inserts when subjected to machining undergoes massive wear at their cutting edges. This worn out inserts result in reduced material removal rate and poor dimensional accuracy. This problem can be solved by using coated inserts and PCD inserts. Even though PCD inserts perform well, due to their high cost, coated inserts are preferred.

Some of the generally used insert coatings are TiAlN, diamond like carbon (DLC), Hyperlox, Tinalox and diamond coatings. Coatings provide high flank wear resistance, which is the common wear that takes place in all metal cutting operations. In case of MMC, the reinforcement wears out the cutting edge by abrasion. This wear mechanism hinders the use of TiAlN coating for MMC as they have less abrasion resistance. Hyperlox coating comprises of a nano-composite structured 2nd generation AlTiN supernitride and it results in a 32 % increase in tool life when compared with TiAlN coating. Figure 6.15 show the uncoated and coated inserts used for conducting the experiments. The general properties of the coating applied in our case are shown in Table 6.11.

At the optimised cutting conditions, experiments are conducted until flank wear reaches 0.25 mm with uncoated inserts and coated inserts separately. The peak flank wear VB_{max} is measured after each run using Digital tool maker's microscope. The results of the experiment are listed in Table 6.12.

The flank wear obtained from the experiments are plotted against the machining time in Fig. 6.16. It is revealed from the Fig. 6.17 that the life of the uncoated insert for 0.25 mm wear is 103 s whereas it is 150 s for coated insert. Hereby, we can observe that the Hyperlox coating results in a 45.6 % increase in tool life.

Table 6.11 Properties of hyperlox

S. No.	Properties	Description
1.	Composition	2nd generation AlTiNsupernitride
2.	Coating structure	Nanocomposite
3.	Colour	Black anthracite
4.	Micro hardness	3,700 VHN
5.	Maximum application temperature	1,100 °C
6.	Coating thickness	3 μm at the nose of insert

Table 6.12 Experimental results under optimised machining conditions

Machining time (s)	Peak flank wear, VB_{max} (μm)	
	Uncoated inserts	Coated inserts
17	43	27
34	89	62
51	136	99
68	166	130
86	197	155
103	253	182
150	–	248

Fig. 6.16 Machining time (vs.) flank wear

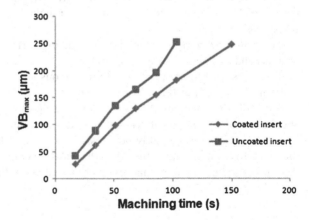

AlTiN coatings with their higher Al content can offer better thermal resistance than TiAlN coatings. Silicon content in the coating which surrounds AlTiN crystallites as a silicon nitride binder ensures that a fine nanostructure is maintained up to 1,200 °C, therefore, the hardness loss at high temperature is minimized. This ensures that coating has sufficient hardness to resist abrasion of the reinforcement. Thus, Hyperlox coating reduces wear caused by both thermal and mechanical loads.

Fig. 6.17 Tool life comparison chart

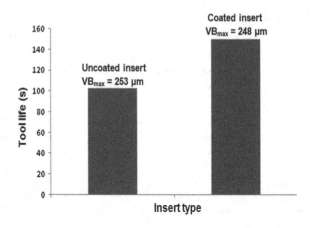

6.3 Summary

The following conclusions are drawn from the experimental analysis:

The tool force is highly affected by the cutting speed followed by feed rate when compared to depth of cut.

The surface finish obtained is better at high speeds and low feed rates. Also depth of cut is found to have a negative effect on the surface finish.

The optimised parameters obtained are 247.83 m/min, 0.05 mm/rev and 0.19 mm for a tool force and surface roughness of 8.74 N and 1.26 μm respectively.

The neural network developed for estimating surface roughness is found to be accurate than response surface method.

The chips formed are generally discontinuous and at high feed rates it contains BuE and Saw tooth at their inner surface.

The tool life is increased by 45.6 % with help of Hyperlox coating and the Al sticking is found be marginally higher in coated inserts which could be because, the Hyperlox coating is highly heat resistant coating. This high heat resistance of the Hyperlox coating melts the Al in the MMC rapidly than the uncoated inserts. Also this can be accounted due the presence of Al in the Hyperlox coating.

References

Arokiadass R, Palaniradja K, Alagumoorthi N (2011) Predictive modeling of surface roughness in end milling of Al/SiCp metal matrix composite. Arch Appl Sci Res 3:228–236
Davim JP (2003) Design of optimisation of cutting parameters for turning metal matrix composites based on the orthogonal arrays. J Mater Process Technol 132:340–344
Ge YF, Xu JH, Yang H, Luo SB, Fu YC (2008) Workpiece surface quality when ultra-precision turning of SiCp/Al composites. J Mater Process Technol 203:166–175

Karakas SM, Acir A, Ubeyli M, Ogel B (2006) Effect of cutting speed on tool performance in milling of B4Cp reinforced aluminum metal matrix composites. J Mater Process Technol 178:241–246

Lin JT, Bhattacharyya D, Fergusod WG (1998) Chip formation in the machining of SiC—particle reinforced aluminium—matrix composites. Compos Sci Technol 58:285–291

Manna A, Bhattacharayya B (2003) A study on machinability of Al /SiC—MMC. J Mater Process Technol 140:711–716

Muthukrishnan N, Paulo Davim J (2009) Optimization of machining parameters of Al/SiC-MMC with ANOVA and ANN analysis. J Mater Process Technol 209:225–232

Oktem H, Erzurumlu T, Kurtaran H (2005) Application of response surface methodology in the optimization of cutting conditions for surface roughness. J Mater Process Technol 170:11–16

Ozben T, Kilickap E, Cakir O (2008) Investigation of mechanical and machinability properties of SiC particle reinforced Al-MMC. J Mater Process Technol 198:220–225

Quan YM, Zhou ZH, Ye BY (1999) Cutting process and chip appearance of aluminium matrix composites reinforced by SiC particle. J Mate Process Technol 91:231–235

Seeman M, Ganesan G, Karthikeyan R, Velayudham A (2010) Study on tool wear and surface roughness in machining of particulate aluminum metal matrix composite-response surface methodology approach. J Adv Manuf Technol 48:613–624

Teti R (2002) Machining of composite materials. University of Naples Federico II, Italy

Printed in the United States
By Bookmasters